REVISE AQA GCSE

Science

Additional Science

REVISION WORKBOOK

Higher

Series Consultant: Harry Smith

Authors: Iain Brand and Mike O'Neill

THE REVISE AQA SERIES

Available in print or online

Online editions for all titles in the Revise AQA series are available Spring 2013.

Presented on our ActiveLearn platform, you can view the full book and customise it by adding notes, comments and weblinks.

Print editions

Additional Science Revision Workbook Higher 9781447942474

Additional Science Revision Guide Higher 9781447942467

Online editions

Additional Science Revision Workbook Higher 9781447942245

Additional Science Revision Guide Higher 9781447942191

Print and online editions are also available for Science A (Foundation and Higher), Additional Science Foundation and Further Additional Science.

This Revision Workbook is designed to complement your classroom and home learning, and to help prepare you for the exam. It does not include all the content and skills needed for the complete course. It is designed to work in combination with Pearson's main AQA GCSE Science 2011 Series.

To find out more visit:
www.pearsonschools.co.uk/aqagcsesciencerevision

Contents

1-to-1 page match with the Additional Higher Guide ISBN 9781447942467

BIOLOGY
1 Animal and plant cells
2 Different kinds of cells
3 Diffusion
4 Organisation of the body
5 Organs and organ systems
6 Plant organs
7 Photosynthesis
8 Limiting factors
9 Biology six mark question 1
10 Distribution of organisms
11 Sampling organisms
12 Transect sampling
13 Proteins
14 Digestive enzymes
15 Microbial enzymes
16 Aerobic respiration
17 The effect of exercise
18 Anaerobic respiration
19 Biology six mark question 2
20 Mitosis
21 Meiosis
22 Stem cells
23 Genes and alleles
24 Genetic diagrams
25 Mendel's work
26 Punnett squares
27 Family trees
28 Embryo screening
29 Fossils
30 Extinction
31 New species
32 Biology six mark question 3

CHEMISTRY
33 Forming ions
34 Ionic comounds
35 Giant ionic structures
36 Covalent bonds in simple molecules
37 Covalent bonds in macromolecules
38 Properties of simple molecules
39 Properties of macromolecules
40 Properties of ionic compounds
41 Metals
42 Polymers
43 Nanoscience
44 Chemistry six mark question 1
45 Atomic structure and isotopes
46 Relative formula mass
47 Paper chromatography
48 Gas chromatography
49 Chemical calculations
50 Reacting mass calculations
51 Reaction yields
52 Reversible reactions
53 Rates of reaction 1
54 Rates of reaction 2
55 Chemistry six mark question 2
56 Energy changes
57 Acids and alkalis
58 Making salts
59 Making soluble salts
60 Making insoluble salts
61 Using electricity
62 Useful substances from electrolysis
63 Electrolysis products
64 Chemistry six mark question 3

PHYSICS
65 Resultant forces
66 Forces and motion
67 Distance–time graphs
68 Acceleration and velocity
69 Forces and braking
70 Falling objects
71 Forces and terminal velocity
72 Elasticity
73 Forces and energy
74 KE and GPE
75 Momentum
76 Physics six mark question 1
77 Static electricity
78 Current and potential difference
79 Circuit diagrams
80 Resistors
81 Series and parallel
82 Variable resistance
83 Using electrical circuits
84 Different currents
85 Three-pin plugs
86 Electrical safety
87 Current and power
88 Physics six mark question 2
89 Atomic structure
90 Background radiation
91 Nuclear reactions
92 Alpha, beta and gamma radiation
93 Half-life
94 Uses and dangers
95 The nuclear model of the atom
96 Nuclear fission
97 Nuclear fusion
98 The lifecycle of stars
99 Physics six mark question 3

100 Biology practice paper
106 Chemistry practice paper
112 Physics practice paper
118 Periodic table
119 Chemistry data sheet
120 Physics equations sheet
121 AQA specification skills
122 Answers
142 Imprint

AQA SKILLS Apply Page 121 — Anything with a sticker like this is helping you to apply your knowledge and practice your skills.

A small bit of small print
Target grade ranges are quoted in this book for some of the questions. Students targeting this grade range should be aiming to get most of the marks available. Students targeting a higher grade range should be aiming to get all of the marks available.

Grade ranges
AQA publishes Sample Assessment Material and the Specification on its website. This is the official content and this book should be used in conjunction with it. The questions in this book have been written to help you practise what you have learned in your revision. Remember: the real exam questions may not look like this.

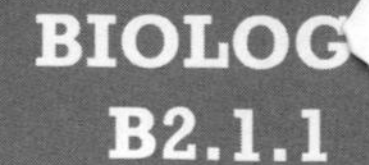

Animal and plant cells

1 The diagram shows a cell from an organism.

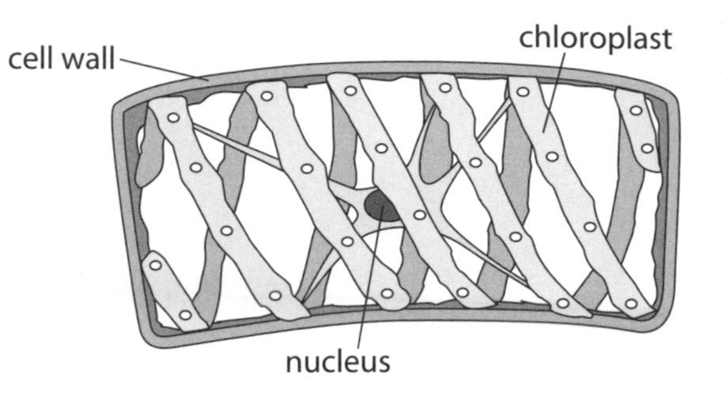

(a) Use information from the diagram to state what type of organism this is.

Plant *(1 mark)*

(b) Identify **two** features that you can see in the diagram that helped you decide on your answer to part **(a)**.

> You need to look at the diagram and name the features that are present in this cell that made you sure of what type of organism this is.

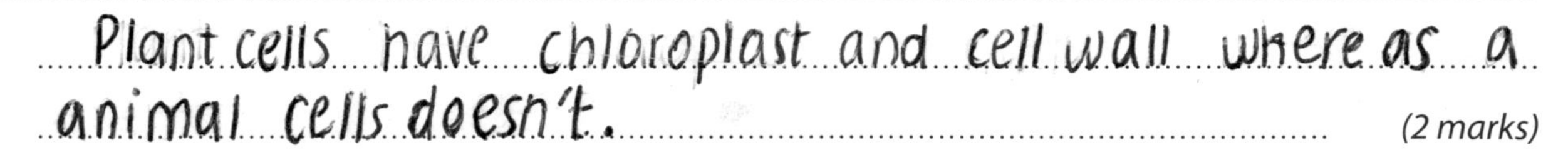

Plant cells have chloroplast and cell wall where as a animal cells doesn't. *(2 marks)*

2 Animal cells contain many different parts.

(a) Describe how a named part of an animal cell is used to release energy.

The mitrocnondria in an animal cell is where energy release occurs.

The energy is released during the process of respiration *(2 marks)*

> On every page you will find a guided question. Guided questions have part of the answer filled in for you to show you how best to answer them.

(b) Give the function of the ribosomes in animal cells.

Proteins are made *(1 mark)*

(c) Plant cells have some parts that are also found in animal cells. Other than those in parts **(a)** and **(b)**, name **one** part that is found inside both plant and animal cells, and give its function.

nucleus; it contains the genes, also it controlls the activities occuring in the cell *(2 marks)*

3 Compare the structure and functions of the cell wall and the cell membrane in plant cells.

AQA SKILL Compare Page 121

Cell membrane controlls what enters and leaves a cell for example: oxygen, whereas the cell wall is strong and is made of cellulose thereby helps maintain the shape of the cell *(4 marks)*

Different kinds of cells

D-C **1** The diagram shows a typical bacterial cell.

DNA
cell membrane
cell wall

(a) Use the diagram to state **one** part of the cell that is common in bacteria and animal cells.

......cell membrane...... *(1 mark)*

Guided **(b)** Use the diagram to describe **one** way in which bacterial cells are different from either plant or animal cells.

Bacterial cells differ from other cells in where their .... is found. In plant and animal cells, this is found in a part of the cell called the ....
In bacterial cells, this is found in .... in the cytoplasm. *(3 marks)*

D-C **2** The diagram shows two specialised cells found in the human body.

AQA SKILL Explain Page 121

cell A
cell B

(a) What type of cell is cell A?

......sperm cell...... *(1 mark)*

(b) Explain how this cell is specialised to carry out its function.

Remember this is an 'explain' question, so you need to identify the feature that makes this cell specialised and say how this is important for the cell's function.

....

.... *(2 marks)*

(c) Cell B is a red blood cell. Give **one** way in which this cell is specialised.

.... *(1 mark)*

(d) The function of this cell is to carry oxygen around the body. How does the specialisation you have identified help the red blood cell to carry out the function of carrying oxygen around the body?

.... *(1 mark)*

D-C **3** Describe the structure of the organism yeast.

....

....

....

.... *(4 marks)*

Diffusion

D-C **1** Many substances move in and out of cells by the process of diffusion.

Guided

(a) What is meant by the term **diffusion**?

Diffusion is the ....spreading.......... of particles, so that there is a net movement from areas of ....high concentration.... to areas of ....lower concentration

(3 marks)

(b) Cells and the fluid surrounding them are mostly water. Suggest **one** property that is needed in substances that diffuse into cells.

.... *(1 mark)*

D-C **2** The diagram shows a microorganism called plasmodium. This microorganism causes the disease malaria.

red blood cell
Plasmodium
10 µm

(a) Use the scale bar to estimate the length of this microorganism.

....

.... *(1 mark)*

(b) Plasmodium needs to obtain a gas by diffusion in order to respire. Name this gas.

........................carbon dioxide........................ *(1 mark)*

(c) Plasmodium gets rid of waste products by diffusion. A student says that diffusion is fast to start with, but stops when the concentrations inside and outside the cell are the same. Do you agree with the student? Explain whether you agree with the statement or not.

Remember that the student's answer may be partially correct. Think very carefully about what happens when the concentrations become the same.

....

.... *(3 marks)*

B-A* **3** The diagram shows the two dissolved substances in neighbouring cells, separated by a cell membrane.

EXAM ALERT

When describing diffusion make sure you understand what **net movement** means.

Students have struggled with questions like this in recent exams – **be prepared!**

substance B
substance A
cell membrane

Explain what happens to these two dissolved substances when the cells are left for some time.

....

....

....

.... *(4 marks)*

Organisation of the body

D-C **1** The diagram shows the respiratory system.
One component of the respiratory system is the lungs.

(a) What level of organisation are the lungs?
Circle the correct answer.

cell	organ	tissue

(1 mark)

nose
windpipe
lung

Guided **(b)** Describe, in terms of levels of organisation, the structure of the lungs.

The lungs are composed of different types of ....

Each of these will be composed of a group of .... with similar structures and functions. *(2 marks)*

(c) Some body systems are used to exchange materials. Give **one** substance that the respiratory system is designed to exchange in the body.

.... *(1 mark)*

D-C **2** Complex organisms such as humans contain many different types of cells. All these different cells arise due to differentiation.

Your answer here needs to say how the cells become different, and why this process occurs.

(a) Describe what is meant by the term **differentiation**.

....

.... *(2 marks)*

(b) One cell that is formed during differentiation is a nerve cell. Nerve cells carry electrical impulses around the body. Describe **one** way in which nerve cells are adapted to perform this function.

nucleus
long fibre

....

....

....

.... *(2 marks)*

B-A* **3** Explain what is meant by the term **tissue**. You should use one example of a tissue found in the body to help you answer the question.

....

....

....

.... *(4 marks)*

Organs and organ systems

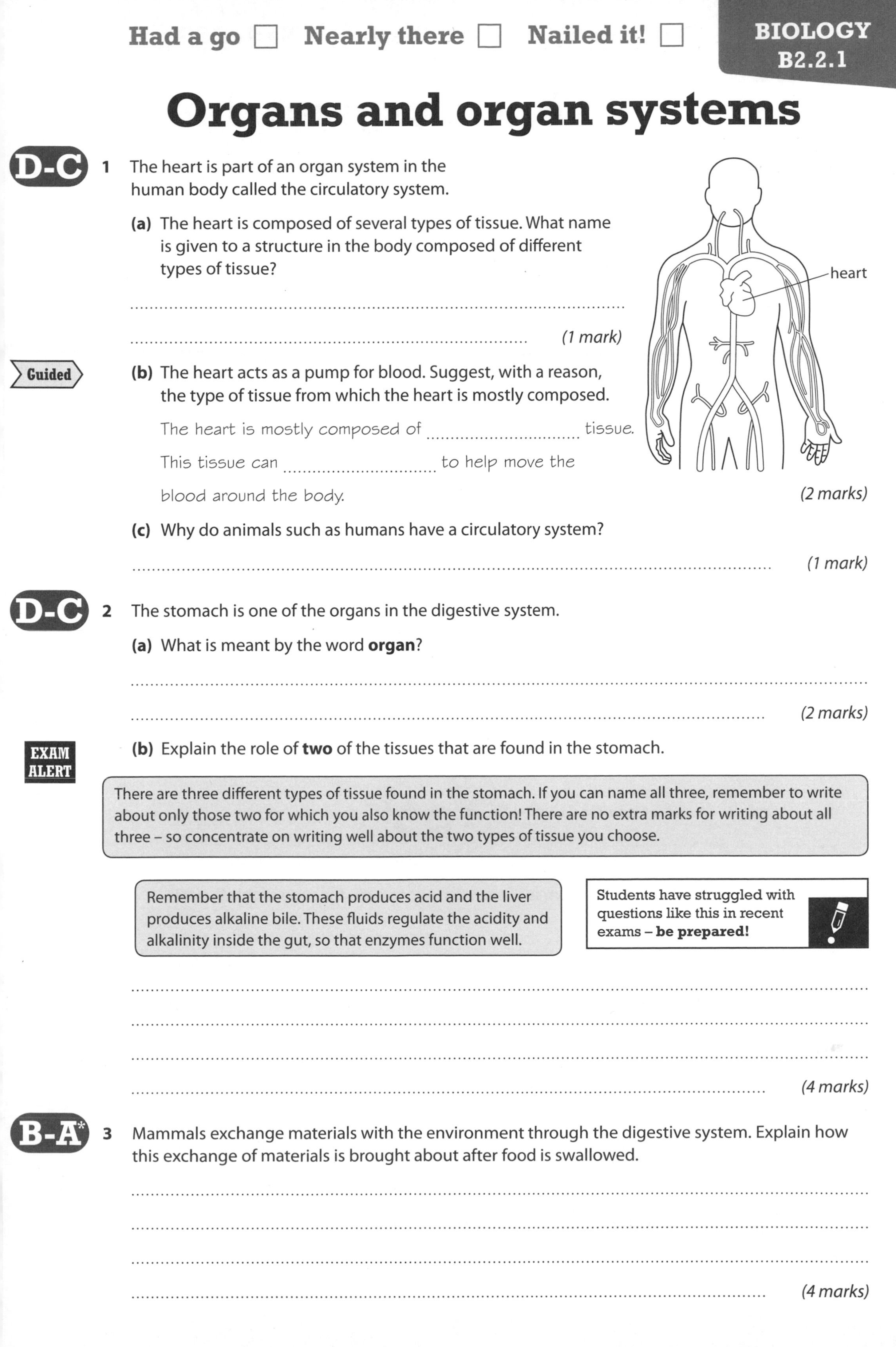

D-C **1** The heart is part of an organ system in the human body called the circulatory system.

(a) The heart is composed of several types of tissue. What name is given to a structure in the body composed of different types of tissue?

....

.... *(1 mark)*

Guided **(b)** The heart acts as a pump for blood. Suggest, with a reason, the type of tissue from which the heart is mostly composed.

The heart is mostly composed of .... tissue.

This tissue can .... to help move the blood around the body. *(2 marks)*

(c) Why do animals such as humans have a circulatory system?

.... *(1 mark)*

D-C **2** The stomach is one of the organs in the digestive system.

(a) What is meant by the word **organ**?

....

.... *(2 marks)*

EXAM ALERT **(b)** Explain the role of **two** of the tissues that are found in the stomach.

There are three different types of tissue found in the stomach. If you can name all three, remember to write about only those two for which you also know the function! There are no extra marks for writing about all three – so concentrate on writing well about the two types of tissue you choose.

Remember that the stomach produces acid and the liver produces alkaline bile. These fluids regulate the acidity and alkalinity inside the gut, so that enzymes function well.

Students have struggled with questions like this in recent exams – **be prepared!**

....

....

....

.... *(4 marks)*

B-A* **3** Mammals exchange materials with the environment through the digestive system. Explain how this exchange of materials is brought about after food is swallowed.

....

....

....

.... *(4 marks)*

Had a go ☐ Nearly there ☐ Nailed it! ☐

Plant organs

D-C **1** Plants have a variety of different organs. These organs include leaves and roots.

Guided **(a)** Describe the function of the leaves of a plant.

The leaves of a plant are adapted to carry out .... in order to provide .... for the plant. *(2 marks)*

(b) Give **two** functions of the roots as an organ of the plant.

....

.... *(2 marks)*

D-C **2** The leaves of a plant are an important organ for the plant. Describe how the composition of a leaf means that it can be described as an organ.

....

....

....

.... *(4 marks)*

B-A* **3** Many plants have stems that are strong and long. The diagram shows a cross-section through the stem of a plant.

B

A

(a) Label A shows the epidermis of the stem. What is the function of this tissue?

.... *(1 mark)*

(b) Use your knowledge to identify and state the function of the tissues labelled B in the diagram.

You may not have seen the cross-section of a stem, but using information in the question you should be able to identify these tissues. As you can see from the diagram, there are two tissues in B – you're not required to know which tissue is which!

....

....

....

.... *(3 marks)*

(c) Suggest **one** other function of the stems of plants.

.... *(1 mark)*

Photosynthesis

D-C

1 Some students investigated how the rate of photosynthesis changed with light intensity. They placed a lamp at different distances from some pond weed and counted the number of bubbles produced by the plant in one minute. The graph shows the results that they collected.

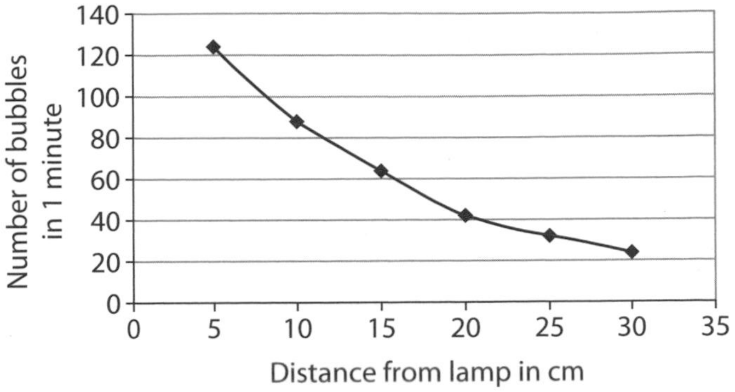

(a) What is the main gas found in the bubbles produced by the plant?

.... *(1 mark)*

Guided

(b) Explain the relationship shown in the graph.

The greater the distance between the lamp and the plant, the .... the light intensity. As light intensity goes ...., the rate of photosynthesis ....

(2 marks)

D-C

2 Plants have many uses for the sugars that are made during photosynthesis.

(a) Some of the sugars are used to make an energy store. Name **one** substance that the plant uses as an energy store.

.... *(1 mark)*

(b) Some substances made from sugars are not used for energy storage. Give the name of one of these products and describe what the plant uses it for.

This question is about a different use of the sugars than the use described in part **(a)**, so make sure you pick a product that is not used for energy storage!

Name: ....

Use: .... *(2 marks)*

B-A*

3 The diagram shows two types of water plant. Water lilies have leaves that float on the surface of the water, whereas pondweed lives under the water. Both plants use their leaves for photosynthesis.

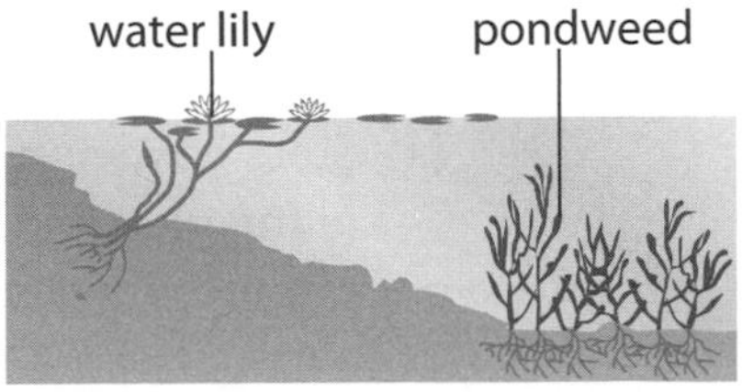

(a) Explain how the plants' leaves are adapted to carry out photosynthesis.

....

....

.... *(3 marks)*

(b) The water lily can take carbon dioxide out of the air and use it for photosynthesis. Explain how the pondweed gets the carbon dioxide it needs for photosynthesis.

....

.... *(2 marks)*

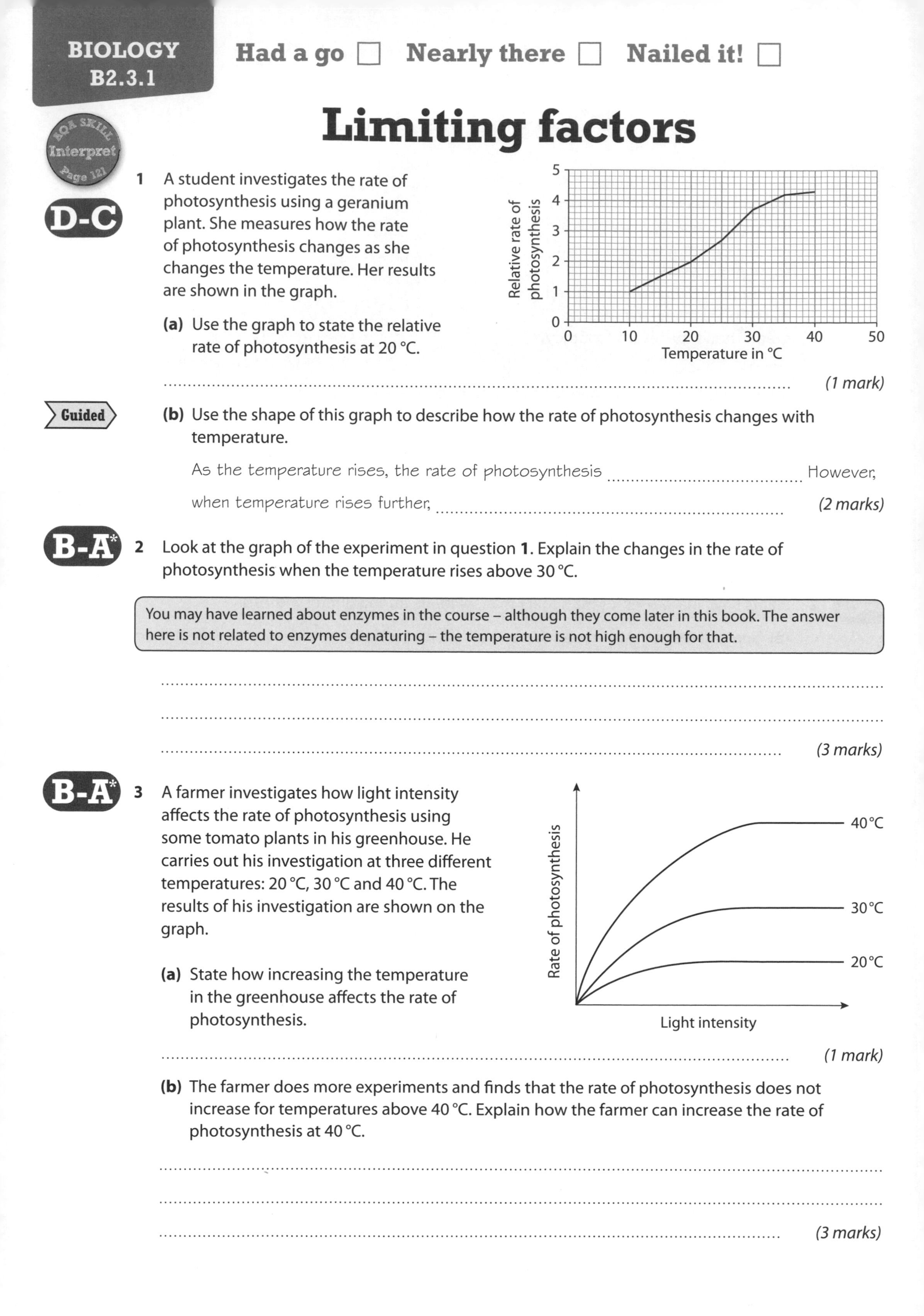

Limiting factors

AQA SKILL Interpret Page 121

D-C

1 A student investigates the rate of photosynthesis using a geranium plant. She measures how the rate of photosynthesis changes as she changes the temperature. Her results are shown in the graph.

(a) Use the graph to state the relative rate of photosynthesis at 20 °C.

.... *(1 mark)*

Guided

(b) Use the shape of this graph to describe how the rate of photosynthesis changes with temperature.

As the temperature rises, the rate of photosynthesis .... However, when temperature rises further, ... *(2 marks)*

B-A*

2 Look at the graph of the experiment in question **1**. Explain the changes in the rate of photosynthesis when the temperature rises above 30 °C.

> You may have learned about enzymes in the course – although they come later in this book. The answer here is not related to enzymes denaturing – the temperature is not high enough for that.

....

....

.... *(3 marks)*

B-A*

3 A farmer investigates how light intensity affects the rate of photosynthesis using some tomato plants in his greenhouse. He carries out his investigation at three different temperatures: 20 °C, 30 °C and 40 °C. The results of his investigation are shown on the graph.

(a) State how increasing the temperature in the greenhouse affects the rate of photosynthesis.

.... *(1 mark)*

(b) The farmer does more experiments and finds that the rate of photosynthesis does not increase for temperatures above 40 °C. Explain how the farmer can increase the rate of photosynthesis at 40 °C.

....

....

.... *(3 marks)*

Biology six mark question 1

Living organisms are made up of cells. Different types of cells have different functions within the organism. Cells are adapted for their functions. Some of them might have specific shapes. Some of them may have different structures inside them.

Some typical cells are shown in the diagram below.

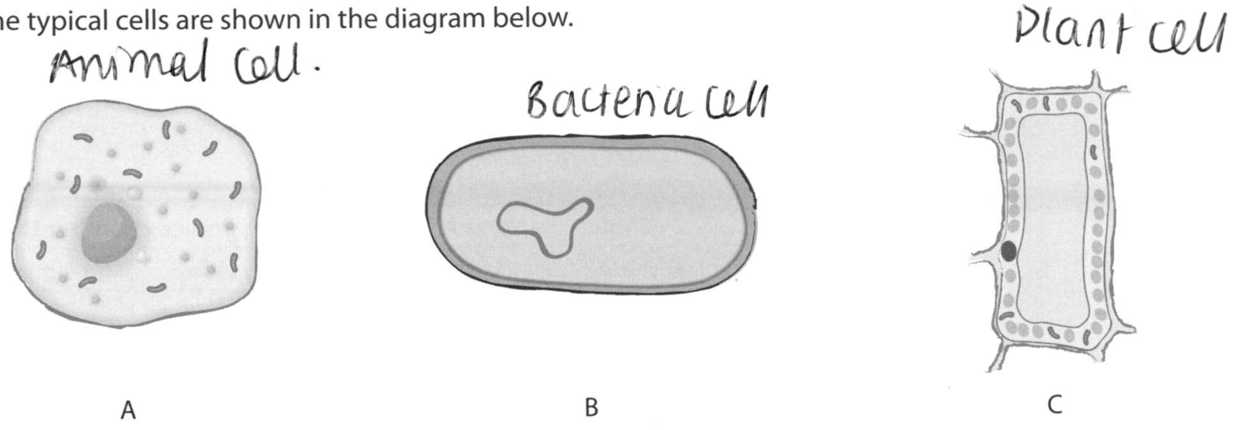

Compare the structure of these cells and the functions of the features found inside them.

> You will be more successful in six mark questions if you plan your answer before you start writing.
> In this question, there are several things that you need to think about:
> - You should compare the cells by referring to the cell structures.
> - For each cell structure that you can identify, you should explain the function that the structure performs.
> - You should use the structures and their functions to identify the types of organism from which the cells might come.
>
> Think about how best to present your answer. One way would be to identify each type of cell and write a few sentences about how the structures it contains led you to that identification.

....

....

....

....

....

....

....

....

....

....

....

....

....

....

....

.... *(6 marks)*

Distribution of organisms

EQA SKILL Suggest Page 121

1 A student carries out an investigation with woodlice to see what conditions they prefer to live in.

She places 10 woodlice on one side of each of the two choice chambers and leaves them for an hour to move around. After this time, she counts the numbers of woodlice on each side of the choice chamber. She repeats the experiment three times and puts her data in a table.

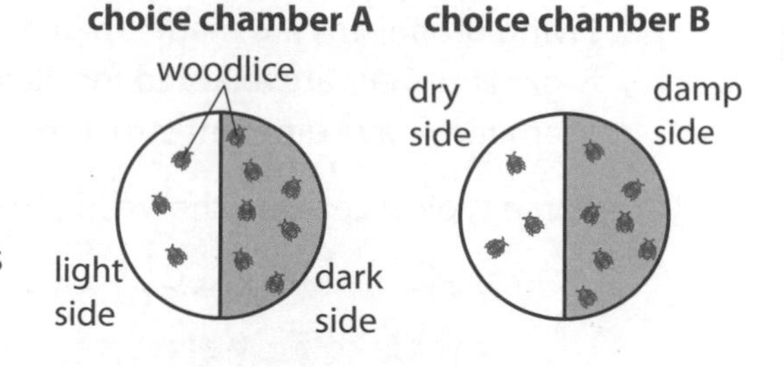

	Choice chamber A		Choice chamber B	
	Dark side	**Light side**	**Dry side**	**Damp side**
Expt 1	8	2	3	7
Expt 2	9	1	2	8
Expt 3	7	3	2	8
Mean	~~7~~ 8	2	2.33	7.67

(a) Complete the table by calculating the mean numbers of woodlice found in each side of choice chamber A. *(2 marks)*

(b) The results of the investigation could have been affected by the method. Explain how the student could make sure that the way she put the woodlice in the choice chamber would not affect the results.

The student would need to put the same amount of woodlice on each side of the chamber at the start of the experiment. This makes sure the experiment is fair, and not biased. *(2 marks)*

(c) Use the results of the student's experiment to suggest why woodlice are often found under rotting tree trunks in the wild.

You need to give a conclusion for the student's experiment for the choice chamber test. Then, relate this to the habitat where woodlice are found.

~~iteu~~ They're found there because woodlice prefer darker sides, and damp areas compared to the dry side and light side. *(3 marks)*

B-A*

EQA SKILL Explain Page 121

2 Lichens are organisms that can grow on the trunks of trees and on other surfaces. Scientists looked at the distribution of lichens in sections of a tree and of a concrete post next to the tree.

Explain the distribution of lichens on the tree and the post.

Percent cover: 0, 10, 20, 30, 40, 50, 60
tree
concrete post
Facing away from sun
Facing sun
Side-on to the sun

There is more lichens facing the sun ~~compared to~~ on the tree than the concrete post compared to when it's away from the sun or if it's side on. There is always more lichens on the tree compared to the concrete post. *(4 marks)*

7 8 8 (11) 12 12 12

Sampling organisms

D-C 1 A gardener goes into his garden every night at 7pm and counts the number of slugs in the same 1 m^2 area of his flower bed. He records his results in a table.

AQA SKILL Describe Page 121

Day	Monday	Tuesday	Wednesday	Thursday	Friday	Saturday	Sunday
Number of slugs	11	12	7	12	8	8	12

(a) Describe how the gardener could make sure the 1 m^2 area of the flower bed was chosen at random on the first day.

~~Repeat the experiment~~ use a 1 m x 1 m quadrat at different places around the flower bed. *(2 marks)*

Guided **(b)** Why does the gardener use a 1 m^2 area each time?

Using the same area means that his experiment is not biased *(1 mark)*

(c) Describe **one** way in which the gardener could improve the reproducibility of the data that he collected.

calculate the mean by taking several repeat samples. *(2 marks)*

D-C 2 A class is investigating the numbers of clover plants on a football pitch. The pitch measures 100 m by 65 m. The class wants to find the total number of clover plants in the field. The teacher gives the class a 1 m by 1 m quadrat.

(a) Explain how the class can use the quadrat to estimate the mean number of clover plants in a 1 m^2 area.

Take quadrat measures around the field and calculate the mean. *(2 marks)*

(b) The class finds that the mean number of clover plants in an area of 1 m by 1 m is 7. Estimate the number of clover plants on the whole football pitch.

100 x 65 = 6500 m^2 x 7 = 45500

(3 marks)

B-A* 3 Use the data in question **1** to explain the difference between the terms mode, mean and median.

In this question, you need to give a definition of each of the terms. To illustrate your definition, you should perform a calculation, using the data in the table, to show each of these forms of average.

Mode: value that is most common which is (12).

Mean: Total number of days divided by amount of slugs added together.

Median: The median is the middle number in order, which is (11) *(6 marks)*

Transect sampling

D-C 1 Limpets are animals that have a shell and live on rocks that are underwater some or all of the time. They can be found in the sea, or in rock pools on the beach. A scientist is investigating the distribution of limpets on the beach.

(a) Explain how the scientist could use a transect to investigate the distribution of limpets.

Place a 1m x 1m quadrats placed at regular intervals along a transect and it will show the distribution of organisms from neighbouring areas. (3 marks)

The scientist sets up three different transects and measures the numbers of limpets on each one. His data is shown in the table.

Distance from sea in metres	Number of limpets			Mean number of limpets
	Transect 1	Transect 2	Transect 3	
0.5	20	23	20	21
1.0	18	16	17	17
1.5	13	13	13	13
2.0	10	8	9	9
2.5	5	6	4	5

(b) Calculate the mean number of limpets at 2.0 m from the sea in this investigation.

10 + 8 + 9 = 27 $\frac{27}{3}$ = 9 (2 marks)

(c) What conclusion can be made from his investigation?

Your conclusion should describe how the distribution of limpets changes along the transect.

As the distance in metres from the sea increases from the transects the no. of limpets decrease as well. (3 marks)

B-A* 2 A scientist investigated the distribution of bluebells in a large wood. She started on the edge of the wood, and measured a line going deeper into the wood. Every 2 m into the woodland, she placed a quadrat and counted the number of bluebells in the quadrat.

Guided

AQA SKILL Describe Page 121

(a) Describe **one** way the scientist could alter her method to collect more accurate data.

The scientist should repeat the counting of bluebells in the quadrat at each distance so that she can calculate the average (2 marks)

(b) The scientist obtained the following data:

Distance from edge of wood in metres	0	2	4	6	8	10	12	14	16
Number of bluebells	0	7	15	22	25	21	16	10	8

Suggest an explanation for these results.

The closer the edge of the wood in metres there is less blue bells from 0-4. There is more bluebels in the middle of the wood from 6-10. As it goes further down the number of bluebells decrease gradually at 12-16 m. (4 marks)

Proteins

D-C

1 Some proteins are biological catalysts in the body.

(a) Give the name for this class of proteins. Tick (✓) **one** box.

☐ antibodies ☑ enzymes ☐ hormones *(1 mark)*

Guided

(b) Explain why this type of molecule is affected by changes in temperature in the body.

EXAM ALERT

Make sure you remember that high temperatures and unsuitable pH levels change the shape of enzymes.

Students have struggled with questions like this in recent exams – **be prepared!**

If it gets too warm, the protein molecule can denature

This changes the shape of the molecule, making it less effective. *(2 marks)*

D-C

2 There are 21 different amino acids that make up proteins in the human body. The body can make some amino acids, but 9 of the amino acids are called essential amino acids and cannot be made by the body.

(a) Suggest where the body gets these essential amino acids.

Antibodies are proteins and they kill pathogens. *(1 mark)*

(b) Proteins in the body have many uses. Explain how some proteins help to prevent illness.

from the diet/food.

(2 marks)

(c) Weightlifters often take protein supplements when they train. Explain how this helps them as they train.

You need to think about the different roles that protein can have in the body. Which of these roles would be useful to a weightlifter in training?

Proteins help the structure in the muscles.

(2 marks)

B-A*

3 Haemoglobin is a protein molecule used to transport oxygen in the blood. The molecule contains four protein units, each with an atom of iron at its centre. The diagram shows a unit of the protein.

protein

iron

Describe how the haemoglobin protein forms.

A long chain of amino acids folds up to form a protein molecule with a 3D shape.

(4 marks)

Digestive enzymes

D-C **1** The stomach churns our food, but also helps in digestion.

(a) Name the enzymes that are active in the stomach.

Proteases .... *(1 mark)*

(b) Describe the function of these enzymes in the stomach.

the enzymes in the stomach ....

....

.... *(3 marks)*

D-C **2** Our food is digested into smaller molecules in the digestive system. Many of the enzymes used perform their functions much more efficiently in alkaline conditions.

Guided **(a)** One organ in the digestive system is the pancreas. Describe the function of the pancreas.

The pancreas produces a variety of .... These are released into a part of the digestive system called .... *(2 marks)*

(b) Fats are one of the types of food molecule. What are the products made when fats are digested?

....

.... *(2 marks)*

(c) After partially digested food leaves the stomach, the body faces a problem in maintaining the correct pH for digestive enzymes to work efficiently. Explain what this problem is, and how the body solves it.

> You should think about what has been happening in the stomach before the partially digested food continues. Your explanation should name the substance the body uses to help solve the problem.

....

....

....

.... *(4 marks)*

B-A* **3** The graph shows how the activity of the enzyme amylase varies with temperature.

Explain why the graph has this shape, in terms of the activity of amylase.

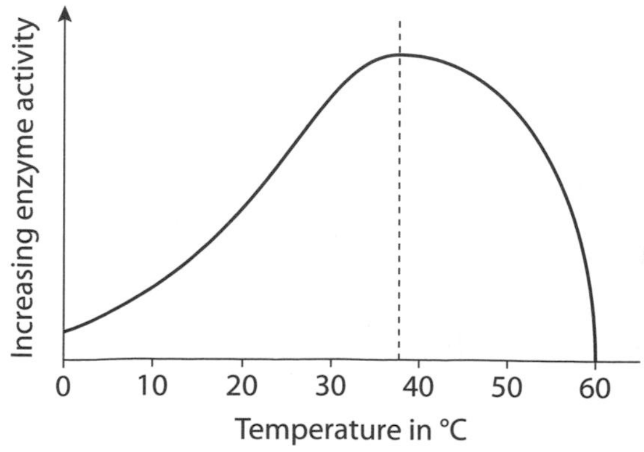

....

....

....

....

....

....

.... *(4 marks)*

Microbial enzymes

D-C **1** Some tests are carried out to find the effectiveness of two biological detergents on some stains on clothes. Two samples of cloth are stained with the same substances. The samples of cloth are washed at 40 °C with two different detergents, A and B.

blood

chocolate olive oil detergent A detergent B

(a) Which enzyme is present in biological detergent A?

.... *(1 mark)*

(b) Biological detergent B contains a protease. What information do these tests give about the composition of blood?

.... *(1 mark)*

Guided **(c)** Use information from these tests to explain why biological detergents often have more than one type of enzyme present.

Some stains may contain more ....

so a single enzyme .... *(2 marks)*

D-C **2** The sweet segments of oranges that are sold in tins are made by placing whole oranges into vats containing protease enzymes. These enzymes break up the peel surrounding the oranges.

(a) Use this information to explain the composition of orange peel.

....

.... *(2 marks)*

(b) Suggest why the oranges are not left in the vats of enzymes for very long periods.

....

.... *(2 marks)*

B-A* **3** Carbohydrase enzymes are one type of enzymes used in the food industry, to convert starch into sugar syrup. Isomerase and protease enzymes are also used in the food industry. Evaluate the use of isomerase and protease enzymes in the food industry.

AQA SKILL Evaluate Page 121

Remember that an evaluation needs to include some advantages and some disadvantages, before coming to a conclusion. There is no need to consider carbohydrase enzymes, or enzymes used outside the food industry.

....

....

....

....

....

....

....

.... *(6 marks)*

Aerobic respiration

D-C **1** Animal cells and plant cells are always respiring.

Guided **(a)** Complete the equation for aerobic respiration.

.... + oxygen → .... + ...

(+ ....) *(2 marks)*

(b) Give the main purpose of respiration in living cells.

.... *(1 mark)*

(c) Suggest why the level of respiration in human cells will change in a 24-hour period.

....

.... *(2 marks)*

D-C **2** Some uses for respiration are the same in plants and animals, and some are different.

(a) Give **two** ways that animals such as mammals use the energy released in respiration, which plants do not.

....

....

.... *(2 marks)*

(b) Describe **one** way that plants use the energy released in respiration for growth.

Your answer here should focus on the way in which plants use respiration to make named 'building blocks' needed to make new material for growth.

....

....

.... *(3 marks)*

B-A* **3** Humans rely on aerobic respiration to survive.

EXAM ALERT

Make sure you use the phrase 'releases energy' when describing respiration – not 'releases heat'.

Students have struggled with questions like this in recent exams – **be prepared!**

(a) Suggest why human muscle cells contain large numbers of mitochondria, but there are fewer mitochondria in human skin cells.

....

....

.... *(2 marks)*

(b) Humans can survive and respire if their body temperature drops by several degrees, but a rise of body temperature of more than a few degrees can be fatal. Suggest why this is the case.

....

....

.... *(2 marks)*

The effect of exercise

D-C

AQA SKILL Interpret Page 121

1 A cyclist wants to investigate how exercise affects his breathing rate. He measures his breathing rate while resting. He then cycles at a constant speed on an exercise bike for 10 minutes. His breathing rate is measured for each minute during the exercise. The graph shows the result of his investigation.

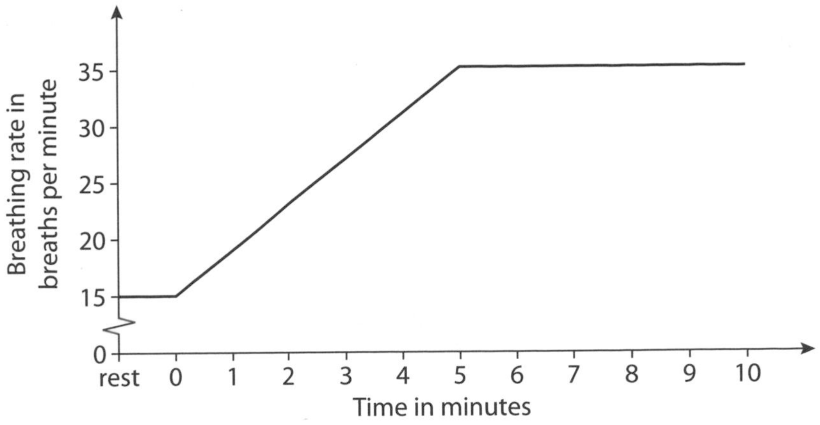

Guided

(a) Describe what the graph shows about the effect of exercise on the cyclist's breathing rate.

The graph shows that exercise .... breathing rate.

However, after a few minutes .... *(2 marks)*

(b) By how much does the cyclist's breathing rate increase during exercise? Give units for your answer.

....

....

Answer .... units .... *(3 marks)*

(c) Before starting exercise, the cyclist was taking a breath every 4 seconds. Show, by calculation, how this gives the initial breathing rate shown on the graph.

....

.... *(2 marks)*

B-A*

2 A group of students measure their heart rate when resting. They then run up and down a flight of stairs and measure their heart rate again.

(a) Explain the pattern the students would see in their data.

....

.... *(2 marks)*

(b) As the students exercise, their muscles use up glucose. Describe the ways in which this glucose is supplied to the students' muscle cells.

> Note that the question asks about the 'ways' in which glucose is supplied – there are two ways in which this happens. You need to think about both routes to score all the marks here.

....

....

....

.... *(4 marks)*

Had a go ☐ Nearly there ☐ Nailed it! ☐

Anaerobic respiration

D-C **1** An athlete is running a long-distance race. Halfway through the race, she notices a pain in the muscles of her side.

(a) What happens to the athlete's muscle cells as she runs the race? Tick (✓) the box next to the correct answer.

☐ They die. ☐ They become fatigued. ☐ They stop respiring. *(1 mark)*

Guided **(b)** Describe how the pain that the athlete feels in her muscles is related to the changes in chemical substances taking place.

A chemical called .... starts to .... in her muscles.

(2 marks)

D-C **2** Humans can respire aerobically or anaerobically. Compare the processes of aerobic and anaerobic respiration.

There are several things that you can compare: for example, the starting materials, the products and when the type of respiration occurs.

....

....

....

.... *(4 marks)*

B-A* **3** The graph shows how oxygen consumption changes before, during and after a period of exercise. Explain the shape of this graph.

AQA SKILLS Explain Page 121

You need to explain the different parts of this curve in terms of the amount of oxygen being used in respiration. Do not forget to explain the shaded area.

EXAM ALERT

Remember the connection between lactic acid production in cells and anaerobic respiration.

Students have struggled with questions like this in recent exams – **be prepared!**

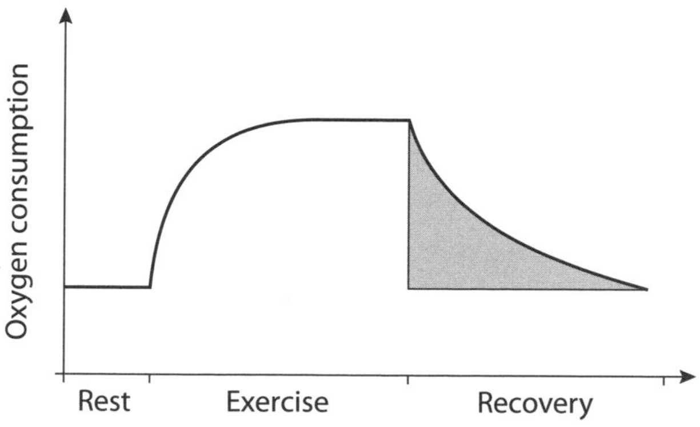

Oxygen consumption

Rest Exercise Recovery

....

....

....

....

....

....

....

.... *(6 marks)*

Biology six mark question 2

Sophie investigates how her heart rate and breathing rate change during exercise. She rests for 2 minutes, then takes light exercise for 5 minutes and heavy exercise for a further 5 minutes.

She takes measurements of her heart rate and breathing rate over this period, and for a further 3 minutes after she finishes the exercise.

The data she collects are shown in the graphs. Explain the shape of these graphs.

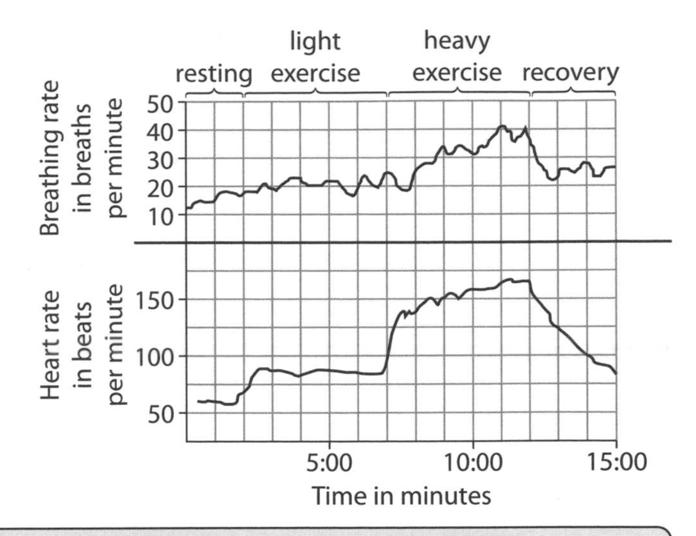

You will be more successful in six mark questions if you plan your answer before you start writing.

One possible approach is to break your answer up to write about the different phases of the investigation the student has carried out. You can explain the change as she moves from one phase to the next, but also the changes that happen within that phase of exercise.

Don't forget to explain what is happening in the recovery phase at the end of the investigation.

Some questions that you might like to think about are:

- Why do the breathing and heart rates change with exercise?
- Why do the breathing and heart rates not continue to increase in an exercise period?
- What happens in cells when exercise continues at the maximum breathing and heart rate?
- Why don't the rates come straight back to the rest rate after exercise stops?

(6 marks)

Mitosis

D-C **1** A cell divides by mitosis every 2 hours.

How many cells are produced from one parent cell in 6 hours?

Guided

At 2 hours: the 1 parent cell divides by mitosis to give 2 daughter cells.

At 4 hours: the 2 cells each divide by mitosis to give .... new cells.

At 6 hours: the .... cells each divide by mitosis to give .... new cells.

(2 marks)

D-C **2** The diagram shows a cell about to divide through mitosis.

(a) How does the genetic composition of the daughter cells compare with the parent cell?

....

....

(1 mark)

(b) In the space below, draw the daughter cells that are produced through mitosis.

Your diagram should show the correct number of chromosomes in each cell.

(2 marks)

B-A* **3** Two types of cell division take place in the body. One type of cell division is meiosis, which produces sex cells. The other type is mitosis.

(a) Give **two** occasions when mitosis takes place in human cells.

....

....

(2 marks)

(b) The graph shows the mass of DNA in the nucleus of a cell over a 24-hour period. Use your knowledge of mitosis to describe the changes occurring in the cell that give the graph this shape.

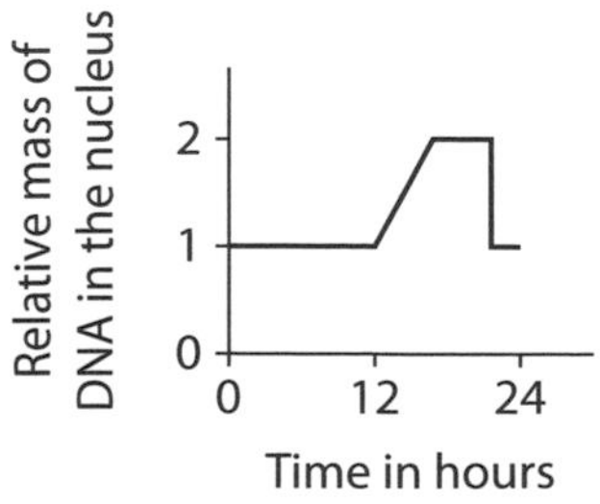

....

....

....

....

....

....

(4 marks)

Meiosis

D-C Guided

1 Human body cells contain our genetic information on the chromosomes.

(a) Describe the number and arrangement of these chromosomes in a human body cell.

Each human body cell contains a total of .... chromosomes.

These chromosomes are normally found in .... in the cells.

(2 marks)

(b) Explain the role chromosomes play in determining the gender of offspring.

....

....

.... *(3 marks)*

D-C

2 The diagram shows a cell about to divide through meiosis.

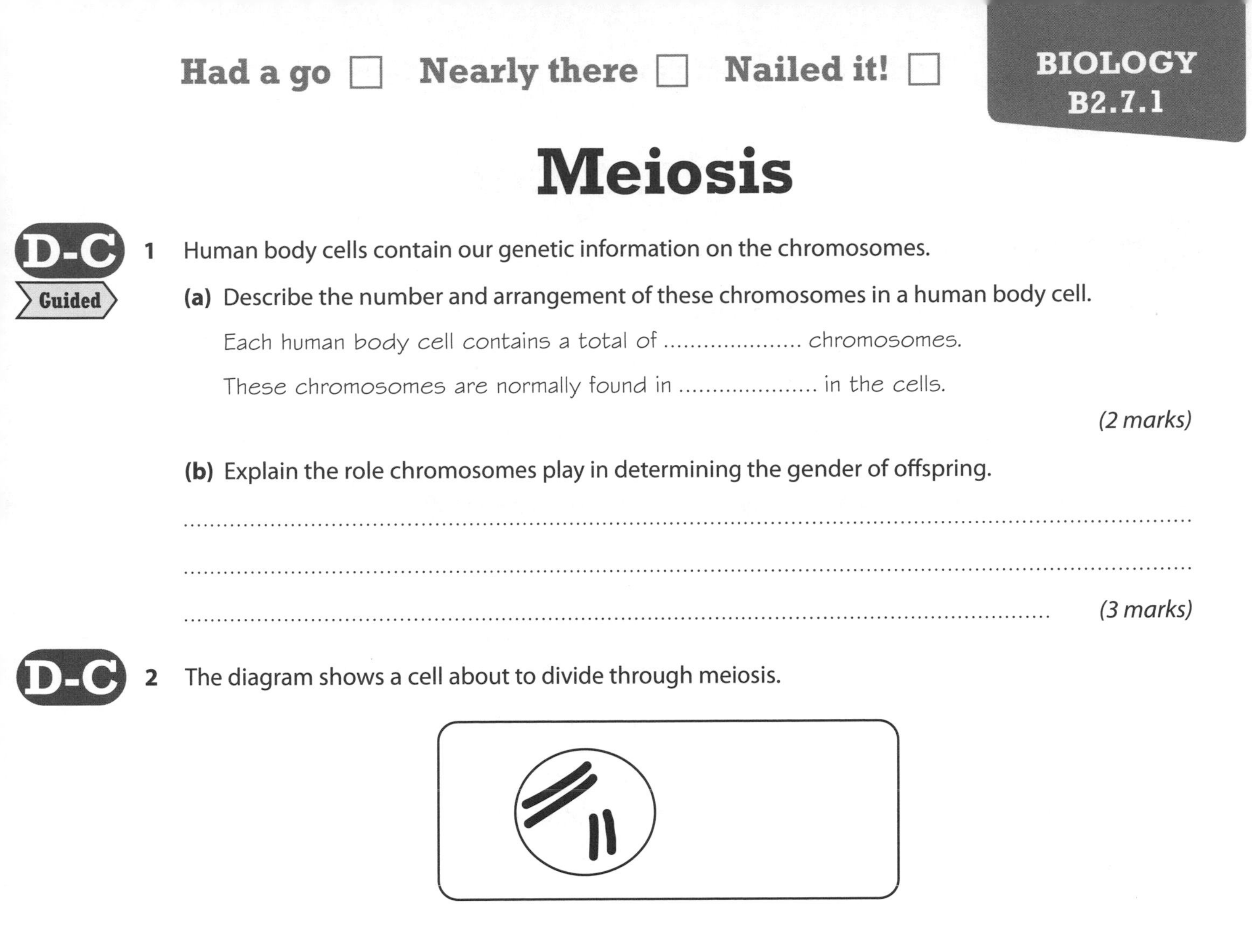

In the space below, draw the daughter cells that are produced through meiosis.

Your diagram should show the correct number of cells, with the correct number of chromosomes in each.

(2 marks)

B-A*

3 A zygote forms as a result of sexual reproduction in humans. This develops into a foetus, which remains in the uterus for nine months, developing into a complex organism.

(a) Describe how the zygote is formed.

....

.... *(2 marks)*

(b) Explain how the process of cell division to make sex cells in the parents differs from the process of cell division that takes place as the foetus develops.

....

....

....

....

.... *(5 marks)*

Had a go ☐ Nearly there ☐ Nailed it! ☐

Stem cells

D-C **1** Stem cells have the ability to differentiate.

Guided **(a)** What is meant by the term **differentiation**?

In differentiation, stem cells become .... to carry out ....

(2 marks)

(b) Compare the way in which plant stem cells and animal stem cells differentiate.

....

....

.... *(3 marks)*

D-C **2** In a recent survey, 1250 people in the UK were asked if they supported embryonic stem cell research. A total of 775 people supported the research; 367 people did not support the research.

(a) How many people in this survey did not give a reply?

.... *(2 marks)*

(b) What percentage of people supported embryonic stem cell research?

....

....

Percentage of people = ... % *(2 marks)*

AQA SKILL Suggest Page 121

(c) Ten years ago, the percentage of people who supported embryonic stem cell research was 26%. Suggest reasons why the percentage has changed over the past ten years.

....

.... *(2 marks)*

B-A* **3** A study in 2003 by scientists in Brazil showed that stem cells could be used to treat paralysis. Stem cells were taken from the blood of the paralysed patients. Each patient received his own stem cells, by injection into the damaged spinal tissues. Out of the 30 patients treated, 12 showed some improvement in their condition. Similar treatments are now being investigated in the USA using embryonic stem cells.

AQA SKILL Evaluate Page 121

Evaluate the potential of these types of stem cell treatments for treating conditions such as paralysis.

> Your answer should mention both types of stem cell treatments. The evaluation should consider negative implications of the treatment as well as positive ones.

....

....

....

....

....

.... *(5 marks)*

Genes and alleles

D-C **1** Our genetic material is stored in chromosomes, which are found in the nucleus of cells. The chromosomes are made from DNA.

Guided **(a)** Describe the structure of the DNA molecule.

DNA consists of two .... of deoxyribonucleic acid, arranged together in a .... *(2 marks)*

(b) Describe the difference between a gene and a chromosome.

....

.... *(2 marks)*

D-C **2** Two parents have three children of different ages. The children have some features in common, but are not identical to each other.

(a) The children each have different alleles. What is an allele?

.... *(1 mark)*

(b) These alleles determine characteristics such as eye colour in the children. Suggest **one** other physical characteristic that alleles may determine in the children.

.... *(1 mark)*

B-A* **3** When humans reproduce sexually, their offspring inherit a combination of genes from both parents.

(a) Genes contain coded information. Describe how the body uses this code.

....

.... *(2 marks)*

(b) A family is made up of a mother and father and identical twins. Explain how the genes that the twins carry compare with their parents and with each other.

Note that this question is an 'explain' question. It's not enough to describe how the genetic composition compares within this family – you must also give reasons. You are not expected to know reasons for the relationship between the twins' genetic composition, but you should know reasons for the relationship between the twins' genetic composition and that of their parents.

....

....

....

.... *(4 marks)*

Genetic diagrams

1 *Drosophila* are fruit flies. Most *Drosophila* have normal wings, but some can have badly formed wings.

Drosophila A

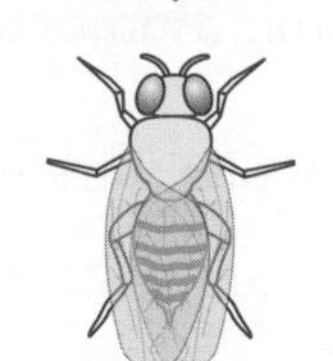

Drosophila B

The alleles for wing type are **D** for normal wings and **d** for badly formed wings. The **d** allele is a recessive allele.

(a) What are the alleles present in the *Drosophila* B in the diagram above? Give a reason for your answer.

....

.... *(2 marks)*

(b) Two *Drosophila* with the alleles **Dd** are mated together to form offspring. The genetic diagram for this cross is shown on the right.

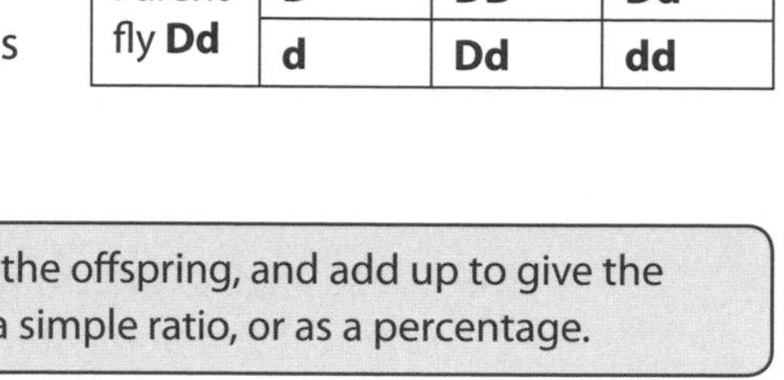

		Parent fly **Dd**	
		D	**d**
Parent fly **Dd**	**D**	**DD**	**Dd**
	d	**Dd**	**dd**

What is the ratio between *Drosophila* with normal wings and those with badly formed wings in the offspring of this mating? Show your working.

> You need to work out the type of wings in each of the *Drosophila* in the offspring, and add up to give the total numbers with each type of wing. You can give your answer as a simple ratio, or as a percentage.

....

....

Ratio = ... *(2 marks)*

B-A*

2 The colour of the fur of a mouse is determined by the genes (**B** and **b**) inherited from the parent mice. The **B** allele gives rise to black fur. Without the **B** allele the mice are brown.

Guided

(a) Explain which genotypes give rise to mice with black fur.

The allele B is .... so mice with the genotypes .... or .... will have black fur. *(3 marks)*

AQA SKILL Predict Page 121

(b) Two mice are bred together. The male mouse is heterozygous for the allele for fur colour. The female mouse is homozygous for this allele, and has brown fur.

(i) What is meant by the terms **heterozygous** and **homozygous**?

....

.... *(1 mark)*

(ii) Use a suitable genetic diagram to work out the phenotypic ratio between the offspring produced when these two mice reproduce.

Ratio = ... *(4 marks)*

Mendel's work

D-C

1 Here are some statements about the work of Mendel. For each statement, explain how this aspect of Mendel's work allowed him to draw valid conclusions from his data.

Guided

(a) Mendel repeated his plant crosses several times.

This allowed Mendel to collect .... of data. This meant that

his results were likely to be .... *(2 marks)*

(b) Mendel pollinated the plants by hand.

....

.... *(2 marks)*

(c) Mendel only used pure-breeding plants.

....

.... *(2 marks)*

D-C

2 After Mendel's initial work, he continued his research by looking at other organisms. However, this type of work has never been done with humans. Explain why neither Mendel, nor other scientists, have repeated this work with humans.

> This question is similar to other ones where you are asked about ethical, social or economic reasons behind scientific issues. You should aim to give two reasons here, with a little explanation. Although you may stick to ethical, social and economic reasons, there are biological reasons you could include here!

....

....

....

.... *(4 marks)*

B-A*

AQA SKILL Explain Page 121

3 One of Mendel's experiments was on flower colour in pea plants. In one case, Mendel found that red-flowered plants crossed with white-flowered plants gave only red-flowered offspring. Modern scientists used Mendel's data to propose the existence of alleles in the pea plants that were responsible for the colour of the flowers. These alleles can be represented as **R** and **r**. The dominant allele is represented by **R.**

(a) Give the three different genotypes for pea plants and the colour of the flower for each genotype.

....

....

.... *(3 marks)*

(b) In one experiment, Mendel bred together pea plants with red flowers that modern scientists would describe as having the genotype **Rr**. Mendel counted the flowers of the offspring of these pea plants and found there were 76 red flowers and 24 white flowers. Explain the data that Mendel obtained.

....

....

.... *(3 marks)*

Had a go ☐ Nearly there ☐ Nailed it! ☐

Punnett squares

D-C

AQA SKILL Predict Page 121

1 A woman with polydactyly marries a man who does not have the condition. She is told by the doctor that her genes are **Pp**. Her husband has the alleles **pp**.
The possible children they could have are shown on the right.

		Mother **Pp**	
		P	**p**
Father **pp**	**p**	**Pp**	**pp**
	p	**Pp**	**pp**

(a) Which allele is responsible for the children of this couple developing polydactyly?

.... *(1 mark)*

Guided

(b) What is the probability that the couple has a child with polydactyly? Give a reason for your answer.

The probability of a child having polydactyly is .... %, because children with the alleles .... show polydactyly, but those with .... not.

(2 marks)

D-C

2 Achondroplasia is a condition where people have shorter than average height. The condition involves the inheritance of the dominant allele, **A**, for achondroplasia. The allele for average height is shown as **a**.

(a) What is meant by the term **dominant allele**?

.... *(1 mark)*

(b) What combination of alleles will be shown by people with average height?

.... *(1 mark)*

(c) A man is tested for his genes and is shown to be **AA**. Explain why any children that this man has will have the condition achondroplasia.

You could answer this question by drawing out different Punnett squares – but you could give a simpler answer by thinking about the alleles passed on by this man and how this would affect the children he has.

....

....

.... *(3 marks)*

B-A*

AQA SKILL Predict Page 121

3 Two parents are homozygous for the alleles for the condition polydactyly – one has the condition and the other does not.

(a) Draw a Punnett square to show the possible genotypes of the offspring of these parents.

Practice drawing Punnett squares.

EXAM ALERT

Students have struggled with questions like this in recent exams – **be prepared!**

(2 marks)

(b) Explain why none of the offspring can be called carriers of polydactyly.

....

.... *(2 marks)*

Family trees

D-C **1** Cystic fibrosis is a recessive genetic disorder.

(a) What does the term **recessive** mean?

.... *(1 mark)*

Guided **(b)** A man is a carrier of cystic fibrosis. Explain what is meant by the term **carrier**, using his alleles as an example.

The man has the alleles .... He is able to ....

for cystic fibrosis, although he does not .... *(3 marks)*

D-C **2** The disease called Batten disease is a recessive genetic disorder.

AQA SKILL Predict Page 121

A man with Batten disease marries a woman who has one allele for the disease and one 'normal' allele. The genetic diagram shows the genetic possibilities for their offspring.

		Mother **Bb**	
		B	**b**
Father **bb**	**b**	**Bb**	**bb**
	b	**Bb**	**bb**

(a) What proportion of the children of this couple are likely to develop Batten disease? Use the genetic diagram to explain your answer.

....

.... *(2 marks)*

(b) Explain how this would change if the mother had the alleles **BB**.

You don't need to draw a genetic diagram or perform any calculations – you can simply think about the alleles being produced by both parents.

....

.... *(2 marks)*

B-A* **3** Sickle cell anaemia is a recessive genetic disorder. People with this condition produce sickle-shaped red blood cells. The recessive allele for sickle cell anaemia can be represented by the letter **h** and the normal allele can be represented as **H**.

AQA SKILL Predict Page 121

Here is a genetic diagram for a couple, neither of whom has sickle cell anaemia.

		Mother **Hh**	
		HH	**h**
Father **Hh**	**H**	**HH**	**Hh**
	h	**Hh**	**hh**

(a) Show that the probability of the couple having a child with sickle cell anaemia is 25%.

....

.... *(2 marks)*

(b) Explain how children of this couple can have sickle cell anaemia, even though neither of the parents exhibit the condition.

....

....

.... *(3 marks)*

Embryo screening

AQA SKILL Explain Page 121

1 One condition for which embryos are screened is Down syndrome. This condition is also known as trisomy-21. People with Down syndrome have three copies of chromosome 21 in their cells.

(a) How many copies of chromosome 21 should there be in our cells?

.... *(1 mark)*

(b) Embryo screening for Down syndrome gives a false positive in 6–7% of cases.

(i) What do you think is meant by the term **false positive**?

.... *(1 mark)*

(ii) Explain why the tests for conditions such as Down syndrome need to be as accurate as possible.

....

.... *(2 marks)*

Guided

(c) The graph shows how the chance of having a child with Down syndrome varies with the age of the mother. Use information in the graph to explain why women of different ages are given different advice on embryo screening for Down syndrome.

The graph shows that the chance of having a child with Down syndrome ....

as the mother's age ....

Therefore, pregnant mothers over the age of 40

....

.... *(2 marks)*

Chance of having child with Down syndrome
9%
7%
5%
3%
1%
20 30 40 50
Maternal age in years

AQA SKILL Evaluate Page 121

B-A* **2** Hereditary fructose intolerance (HFI) is a recessive genetic condition. People with HFI lack an enzyme that breaks up fructose. Symptoms of the condition can include vomiting, fainting and, in the long term, kidney failure. People with HFI avoid foods containing fructose, such as fruits, in their diet.

(a) Evaluate the use of embryo screening to detect HFI.

Your answer should consider the nature of the condition and the reasons – economic, ethical and social – in testing for the condition, as well as what action parents may take as a result of the test.

....

....

....

....

.... *(4 marks)*

(b) Why would people with HFI find it difficult to lose weight using slimming foods?

....

.... *(2 marks)*

Fossils

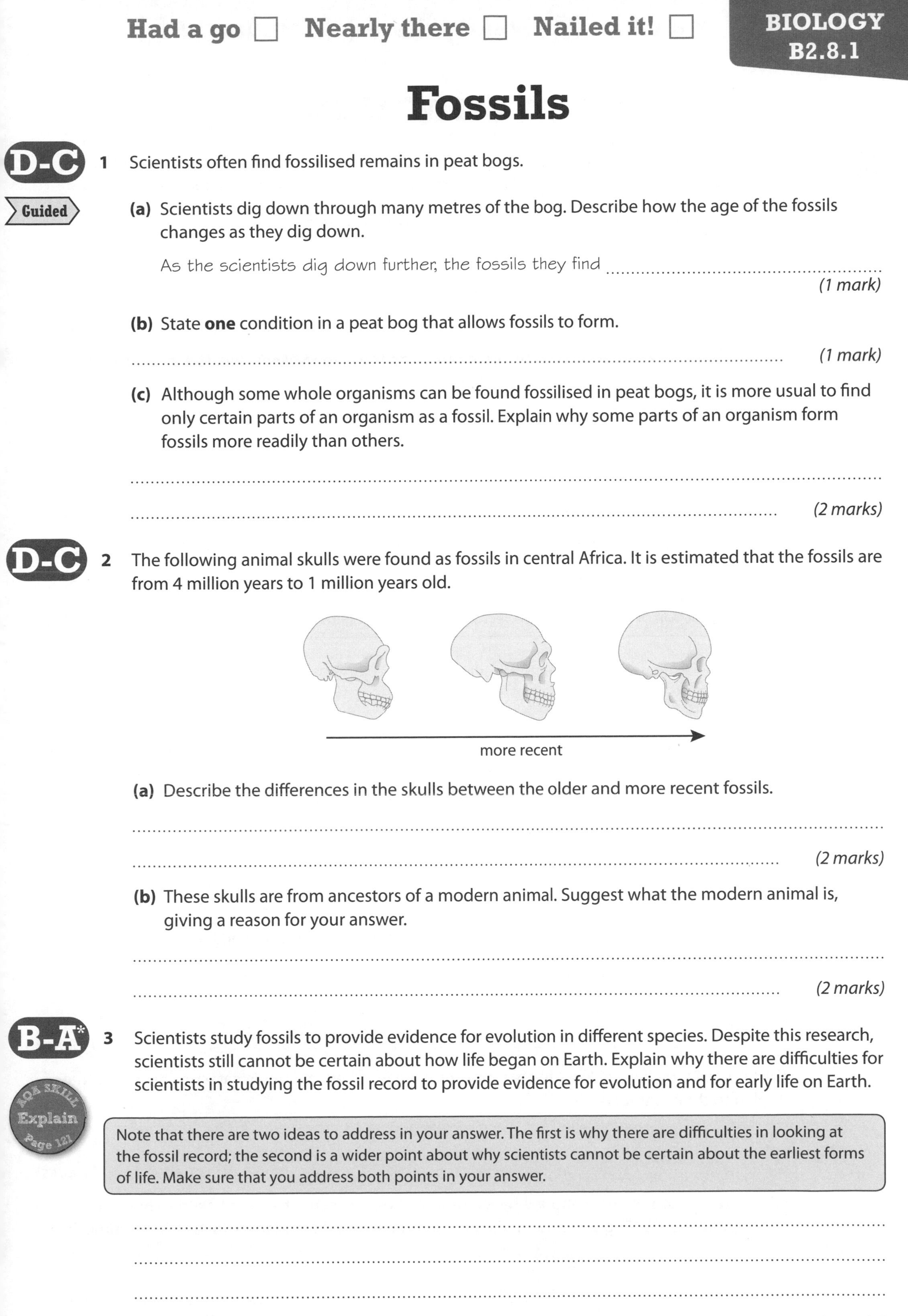

D-C **1** Scientists often find fossilised remains in peat bogs.

Guided

(a) Scientists dig down through many metres of the bog. Describe how the age of the fossils changes as they dig down.

As the scientists dig down further, the fossils they find ....

(1 mark)

(b) State **one** condition in a peat bog that allows fossils to form.

.... *(1 mark)*

(c) Although some whole organisms can be found fossilised in peat bogs, it is more usual to find only certain parts of an organism as a fossil. Explain why some parts of an organism form fossils more readily than others.

....

.... *(2 marks)*

D-C **2** The following animal skulls were found as fossils in central Africa. It is estimated that the fossils are from 4 million years to 1 million years old.

(a) Describe the differences in the skulls between the older and more recent fossils.

....

.... *(2 marks)*

(b) These skulls are from ancestors of a modern animal. Suggest what the modern animal is, giving a reason for your answer.

....

.... *(2 marks)*

B-A* **3** Scientists study fossils to provide evidence for evolution in different species. Despite this research, scientists still cannot be certain about how life began on Earth. Explain why there are difficulties for scientists in studying the fossil record to provide evidence for evolution and for early life on Earth.

AQA SKILL Explain Page 121

Note that there are two ideas to address in your answer. The first is why there are difficulties in looking at the fossil record; the second is a wider point about why scientists cannot be certain about the earliest forms of life. Make sure that you address both points in your answer.

....

....

....

.... *(4 marks)*

Had a go ☐ Nearly there ☐ Nailed it! ☐

Extinction

D-C **1** Christmas Island is an isolated island in the Pacific Ocean. When humans first settled Christmas Island in the 1890s, scientists noted two species of rat that were unique to the island.

(a) Explain why the Christmas Island rats were not seen elsewhere.

....

....

.... *(3 marks)*

(b) In 1899, black rats arrived on the island through the ship SS *Hindustan*. These rats carried a parasite called a trypanosome. In 1904, scientists found that the number of native rats was small; and by 1908, there were no native rats left on the island.

Guided

(i) How long did it take for the Christmas Island rats to become extinct?

The rats were extinct in 1908, after the introduction of black rats in 1899 – this is

.... years. *(1 mark)*

(ii) Describe how the introduction of black rats led to the extinction of the native rats.

You need to use information in the question to help answer this, as there are potentially different reasons why the introduction of black rats could affect the population of native rats.

....

.... *(2 marks)*

B-A* **2** Dinosaurs are thought to have become extinct around 65 million years ago, whereas mammoths became extinct about 8000 years ago. Early humans began to appear around 2 million years ago, evolving into modern *Homo sapiens* around 300 000 years ago.

(a) The extinction of the dinosaurs is often referred to as a mass extinction.

(i) What is meant by the term **mass extinction**?

.... *(1 mark)*

(ii) Suggest how a single catastrophic event can lead to a mass extinction.

....

....

.... *(3 marks)*

(b) Suggest why it is unlikely that mammoths were involved in a single catastrophic event that led to their extinction.

....

.... *(2 marks)*

(c) There are two theories about why mammoths became extinct. One is linked to human population rise. Suggest how an increase in humans could lead to extinction in mammoths.

....

.... *(2 marks)*

New species

D-C **1** The porkfish is a fish that lives in the seas between North and South America. Millions of years ago, land started to emerge in this sea, finishing with the formation of Panama around 3.5 million years ago.

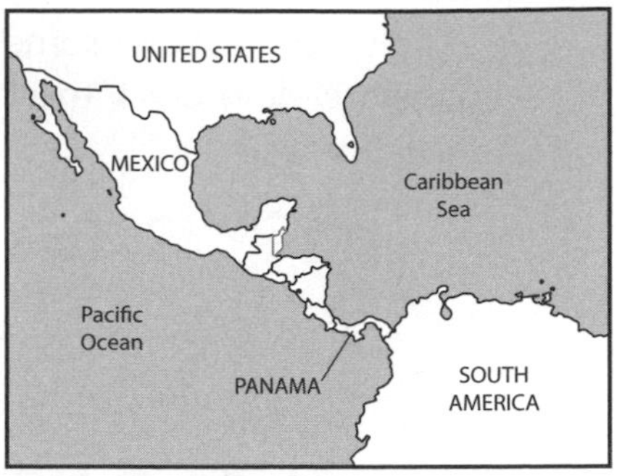

EXAM ALERT

Practice writing about the formation of a new species and make sure you get all the stages in the right order.

Students have struggled with questions like this in recent exams – **be prepared!**

UNITED STATES
MEXICO
Caribbean Sea
Pacific Ocean
PANAMA
SOUTH AMERICA

(a) Describe why the formation of Panama led to changes in the population of porkfish.

....

.... *(2 marks)*

Guided

(b) The modern porkfish has developed through speciation into two different fish. One of these lives in the Caribbean Sea and the other in the Pacific Ocean. Explain what is meant by **speciation**.

Speciation is the development of ....

that cannot .... *(2 marks)*

D-C **2** Darwin described different finch species on the Galapagos Islands. Describe how geographical isolation on the Galapagos led to the development of the different species of finches.

....

....

....

.... *(4 marks)*

B-A* **3** Salamanders are amphibians that look like lizards. Different salamander species of the genus *Ensatina* can be found around the Central Valley of California.

Each species of salamanders is able to breed with its neighbours, with one exception. In Cuyamaca State Park, the two species of salamander *E. eschscholtzi* and *E. klauberi* cannot interbreed.

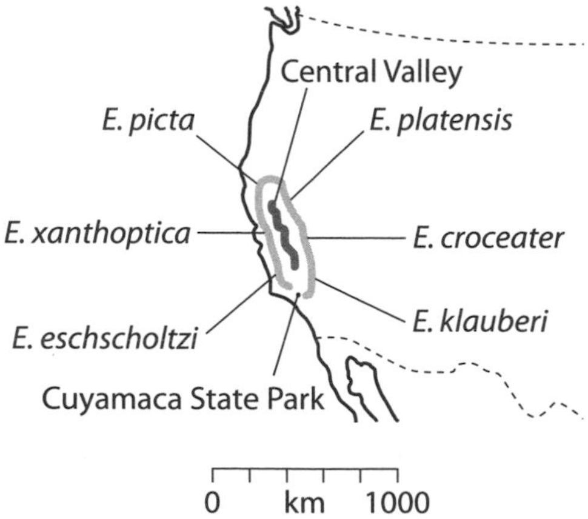

Central Valley
E. picta
E. platensis
E. xanthoptica
E. croceater
E. eschscholtzi
E. klauberi
Cuyamaca State Park
0 km 1000

(a) What term is given to the development of species, such as *E. eschscholtzi* and *E. klauberi*, that cannot interbreed?

.... *(1 mark)*

(b) Suggest an explanation for why these six different salamander species have arisen here in California.

....

....

....

.... *(4 marks)*

Biology six mark question 3

The diagram shows some stages in sexual reproduction, where the single cell formed in fertilisation (zygote) develops into a ball of cells (embryo) and then an organism with several different types of cell (foetus).

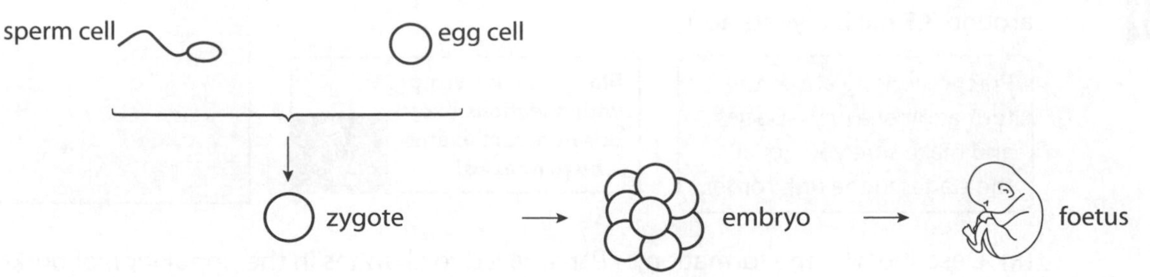

Use the diagram, and your own knowledge of cell division and differentiation, to describe the cell processes that take place at each stage of this process.

You will be more successful in six mark questions if you plan your answer before you start writing.

In this question, there are three different processes taking place – with each process being shown by the arrow linking the different stages. Your answer should look at each stage and should describe:

- how the number of cells present changes at each stage
- the process being used to change the number of cells
- whether there are any changes to the types of cells present, and how these changes are brought about.

Don't forget that the best answers will use technical words correctly. Here, there are many different technical words that you can use:

mitosis	**meiosis**	**fertilisation**	**gamete**	**stem cell**	**differentiation**

Make sure that you use them correctly!

....

....

....

....

....

....

....

....

....

....

....

....

....

....

....

.... *(6 marks)*

Forming ions

EXAM ALERT

1 Complete the table with the missing symbols.

Element	potassium	calcium	oxygen
Symbol	K		
Ion	K^+		

(2 marks)

You may need to refer to the periodic table on page 118.

In an exam you will be given a data sheet and periodic table. In this book they are found on pages 118–119.

Students have struggled with questions like this in recent exams – **be prepared!**

D-C **2** The structures of four particles are described below:

Particle A has 9 protons, 9 neutrons and 9 electrons.

Particle B has 9 protons, 9 neutrons and 10 electrons.

Particle C has 10 protons, 9 neutrons and 10 electrons.

Particle D has 10 protons, 10 neutrons and 10 electrons.

(a) Which of the particles is a negative ion? .... *(1 mark)*

(b) Which of the particles is a fluorine atom? .... *(1 mark)*

B-A* **3** When the atoms of different **elements** join together a **compound** is formed. While there are millions of different compounds, there are in general only two ways that atoms within them form a **bond**.

Guided

(a) Write down a meaning for each of the terms in bold in the above passage.

Element: ....

Compound: ....

Bond: A force of attraction that .... *(3 marks)*

(b) Approximately how many elements are there? .... *(1 mark)*

(c) Name the **two** main types of bond and briefly describe how each is formed.

....

....

.... *(4 marks)*

B-A* **4** The electronic structure of a magnesium atom is shown on the right.

(a) The symbol for a magnesium atom is Mg. Write down the symbol for a magnesium ion.

.... *(1 mark)*

2+

(b) Explain in terms of electrons and electron arrangements what happens when a magnesium atom forms a magnesium ion.

....

.... *(2 marks)*

(c) Name another element in the first 20 in the periodic table that would form an ion in the same way as magnesium. .... *(1 mark)*

Ionic compounds

D-C **1** Sodium chloride is a typical ionic compound with a lattice structure.

(a) Describe a lattice structure.

....

....

.... *(2 marks)*

On every page you will find a guided question. Guided questions have part of the answer filled in for you to show you how best to answer them.

(b) Which **two** kinds of elements usually form ionic compounds?

.... *(1 mark)*

Guided **(c)** Describe what happens to the atoms of sodium and chlorine when they react to make them form a lattice structure.

Sodium atoms each .... an electron to form .... ions and

chlorine atoms each ....

The positive and ....

to form the lattice structure. *(3 marks)*

D-C **2** Some common ions are given in the table on the right.

Use the information in the table to write the formulae for the following compounds.

AQA SKILL Formulae Page 121

(a) zinc bromide .... *(1 mark)*

(b) zinc sulfide .... *(1 mark)*

(c) potassium nitride .... *(1 mark)*

Element	Ion symbol
potassium	K^+
zinc	Zn^{2+}
bromine	Br^-
sulfur	S^{2-}
nitrogen	N^{3-}

B-A* **3** Sodium reacts violently with fluorine to form sodium fluoride.

(a) Name the group of elements in the periodic table to which sodium belongs.

.... *(1 mark)*

(b) Complete the balanced equation for the formation of sodium fluoride.

2Na + ... → 2 .... *(2 marks)*

(c) Fluorine reacts with most metals. The formula of titanium fluoride is TiF_3.
What is the charge on the titanium ion? .... *(1 mark)*

B-A* **4** Chlorine and fluorine form ionic compounds with metals.

(a) Name the group in the periodic table to which chlorine and fluorine belong.

.... *(1 mark)*

(b) Explain, in terms of electronic structure, why chlorine atoms would not form an ionic compound with fluorine atoms.

....

....

.... *(3 marks)*

Giant ionic structures

1 The diagrams below represent an atom of calcium and an atom of oxygen.

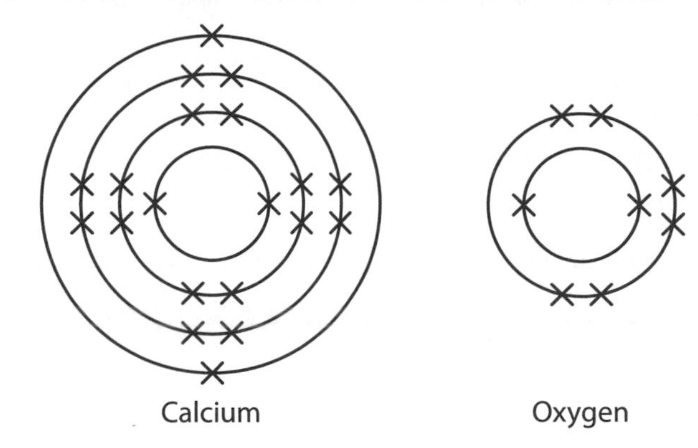

(a) Draw arrows on the above diagrams to show what happens to the electrons when calcium and oxygen form an ionic bond. *(1 mark)*

Guided

(b) Describe the changes that occur to the electronic structure of calcium atoms when they form ionic bonds.

The calcium atoms lose .... which gives them

.... like a noble gas. *(2 marks)*

(c) Write down the symbols for the **two** ions formed when calcium and oxygen form an ionic bond. .... *(2 marks)*

Remember that when the main group elements form bonds, their atoms try to get an electron arrangement like the nearest noble gas.

D-C

2 The diagram on the left represents a sodium atom. Complete the diagram on the right to show the electronic structure and ion charge in a sodium ion.

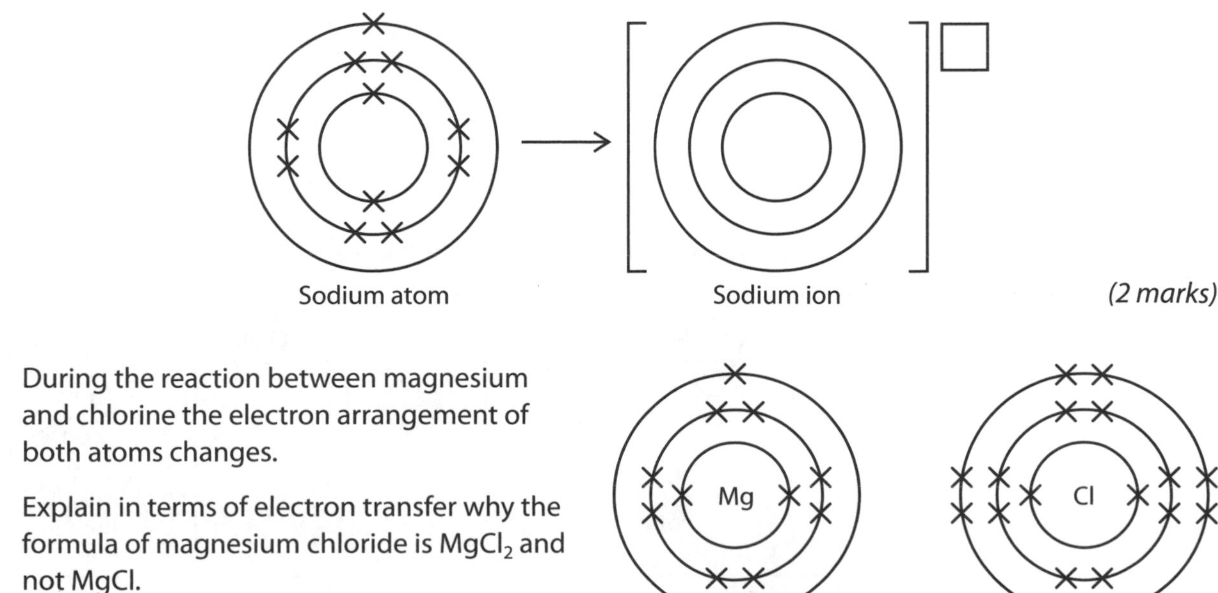

(2 marks)

B-A*

3 During the reaction between magnesium and chlorine the electron arrangement of both atoms changes.

Explain in terms of electron transfer why the formula of magnesium chloride is $MgCl_2$ and not MgCl.

....

....

.... *(3 marks)*

Had a go ☐ Nearly there ☐ Nailed it! ☐

Covalent bonds in simple molecules

D-C 1 Some elements and compounds, like oxygen (O_2), hydrogen chloride (HCl), and methane (CH_4), have covalent bonds between the atoms. These elements and compounds usually have a molecular structure, although there are a few giant covalent structures that are held together by covalent bonds.

(a) What kind of elements usually form covalent bonds? ....

(1 mark)

Guided **(b)** Describe the difference between a molecular structure and a giant covalent structure.

Molecules are small groups of atoms held together by covalent bonds. Giant covalent structures .... *(2 marks)*

D-C 2 Carbon and hydrogen atoms react to form a molecular structure. The bonds in the molecules are formed by the interaction of the electrons in the atoms.

The diagrams on the right show how the electrons are arranged in carbon and hydrogen atoms.

Carbon Hydrogen

(a) Name and give the formula for the compound formed between carbon and hydrogen.

.... *(2 marks)*

(b) **(i)** What type of bonds are formed between carbon and hydrogen atoms?

.... *(1 mark)*

(ii) Describe how these bonds are formed. ....

.... *(1 mark)*

(c) Describe a molecular structure.

....

.... *(2 marks)*

B-A* 3 The diagrams represent the structure of three different covalent substances.

Diagram of molecules			
Name of compound		hydrogen chloride	
Formula		HCl	

The name and formula has been given for one of the compounds. Suggest a possible name and formula for the other two compounds. *(2 marks)*

Covalent bonds in macromolecules

D-C

Guided

AQA SKILL Draw Page 121

1 The structural formulae for hydrogen and ammonia are shown.

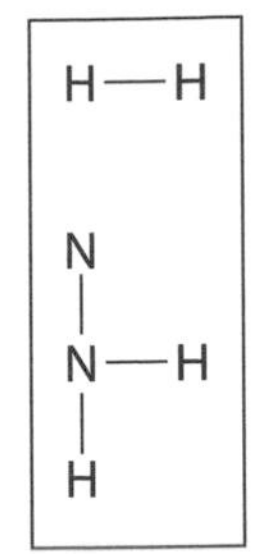

(a) (i) What do the lines between the atoms in the structural formulae represent?

The lines represent shared pairs of ....

which form the .... bonds. *(2 marks)*

(ii) Draw a similar structural formula for silicon hydride, formula SiH_4.

(1 mark)

(b) The bonding in hydrogen molecules can also be represented using a diagram that shows the electron energy levels, like the one shown.

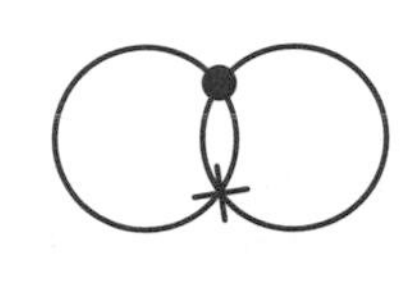

Draw a similar diagram for hydrogen chloride. You need only show the electrons in the highest energy level.

(2 marks)

D-C

2 The diagram on the right represents the outer electrons in the atoms in a nitrogen molecule.

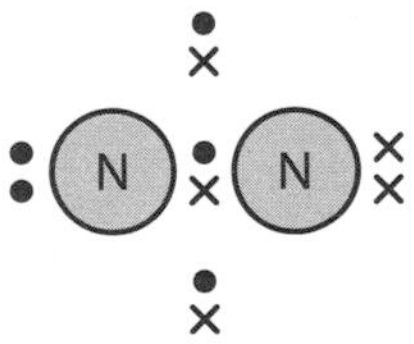

N N

(a) What is the formula for a molecule of nitrogen?

.... *(1 mark)*

(b) How many bonds are there between the nitrogen atoms? .... *(1 mark)*

B-A*

AQA SKILL Draw Page 121

3 The dot and cross diagram for a molecule of chlorine is shown on the right. Draw a similar diagram to show how the outer electrons are arranged in ammonia, NH_3.

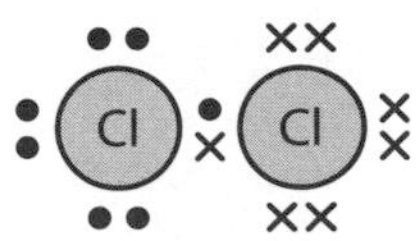

Cl Cl

(2 marks)

Properties of simple molecules

D-C

1 The diagram on the right shows a few molecules of a gas similar to oxygen.

weak forces

strong forces

(a) Draw arrows on the diagram to show where the strong covalent bonds and weak forces of attraction exist. *(1 mark)*

(b) Explain why this substance has a low boiling point even though the covalent bonds that hold its molecules together are strong.

....

....

.... *(3 marks)*

Guided

(c) Explain why this substance does not conduct electricity.

Freely moving charges are needed to conduct electricity and ....

.... *(2 marks)*

B-A*

2 (a) A student was given the following apparatus and asked to test solutions to find out if they were conductors or insulators.

two carbon rods	**beaker**	**light bulb**	**connecting wires**
power supply	**switch**	**test solutions**	

Draw a labelled diagram to show how the apparatus should be set up to test the solutions.

Make sure you label all the given apparatus.

(4 marks)

(b) The following results were obtained by testing aqueous solutions of substances A, B, C and D.

Solution	A	B	C	D
Bulb lit?	Yes	No	Yes	Yes

What conclusions can be drawn from these results?

....

.... *(2 marks)*

Properties of macromolecules

1 The following elements and compounds all contain covalent bonds.

carbon dioxide	diamond	graphite	hydrogen chloride	silicon dioxide	water

The table below splits these substances into two groups depending on their bonding and structure. Complete the table, including suitable headings for each of the groups.

Small molecules	
carbon dioxide	diamond
water	

(6 marks)

B-A* 2 The diagram shows some of the atoms in a piece of graphite.

(a) Describe the **two** different forces of attraction in this structure.

....

....

.... *(2 marks)*

(b) Explain why graphite feels slippery and can be used as a lubricant.

....

.... *(2 marks)*

AQA SKILL Explain Page 121

(c) Explain why graphite can conduct electricity but diamond cannot.

....

....

.... *(3 marks)*

(d) Describe **one** similarity and **one** difference between graphite and a fullerene structure.

....

....

.... *(3 marks)*

B-A* 3 The atoms of both carbon dioxide and silicon dioxide are held together by the same kind of bonds.

(a) What kind of bonding is involved in carbon dioxide and silicon dioxide? ....
(1 mark)

(b) Explain, in terms of the forces involved, why carbon dioxide is a gas at room temperature while silicon dioxide is solid with a high melting point.

....

....

....

.... *(4 marks)*

Had a go ☐ Nearly there ☐ Nailed it! ☐

Properties of ionic compounds

D-C **1** Compound X has a structure similar to the one shown in the diagram.

(a) Name the type of bonding in compound X.

.... *(1 mark)*

(b) Explain what forces of attraction hold this structure together.

....

.... *(2 marks)*

Guided **(c)** The conductivity properties of compound X change when it dissolves in water. Explain the difference in conductivity between solid X and a solution of X.

Solid X does not conduct electricity because the ions ....

A solution of X ....

.... *(3 marks)*

D-C **2** Sodium chloride is the main compound in common salt that is used in cooking. It is a white crystalline solid with a high melting point.

(a) Write a balanced equation for the formation of sodium chloride from its elements.

.... + ... → .... *(2 marks)*

Remember that chlorine atoms form molecules.

(b) Explain why sodium chloride has a high melting point.

....

.... *(2 marks)*

D-C **3** Barium reacts quickly when heated in chlorine gas, to produce white crystals of barium chloride. Barium chloride is an ionic compound with a melting point of 963 °C.

(a) Write the formula for barium chloride. .... *(1 mark)*

(b) What happens to the barium and chlorine atoms to form the ionic bonds?

....

.... *(2 marks)*

(c) Describe the structure of an ionic compound like barium chloride, and the forces that hold it together.

....

.... *(3 marks)*

B-A* **4** Lithium chloride has a similar structure to sodium chloride. Use the structure of lithium chloride to explain its melting point and conductivity.

....

....

....

.... *(4 marks)*

Metals

D-C **1** In recent years scientists have found several medical uses the for new shape memory alloys.

(a) What special property does a shape memory alloy have?

Guided

A shape memory alloy can return to ....

.... *(2 marks)*

(b) Name **one** example of a shape memory alloy and describe **one** use for the metal.

....

.... *(2 marks)*

B-A* **2** Aluminium wire is used in electrical power cables. The graph shows how the conductivity of aluminium changes with temperature.

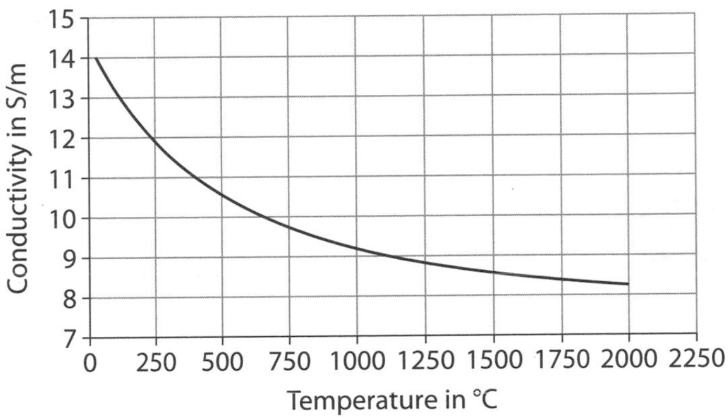

(a) Describe how metals like aluminium are able to conduct electricity.

....

....

.... *(2 marks)*

(b) Describe the trend in conductivity over the range of temperatures used in the graph.

....

.... *(2 marks)*

(c) Aluminium is used for power cables because it conducts electricity. Suggest **two** other properties of aluminium that make it useful for making the cables for overhead electrical power lines.

....

.... *(2 marks)*

B-A* **3** Because of their range of properties, alloys find many uses in industry and the home. The diagram shows a typical arrangement of atoms in an alloy.

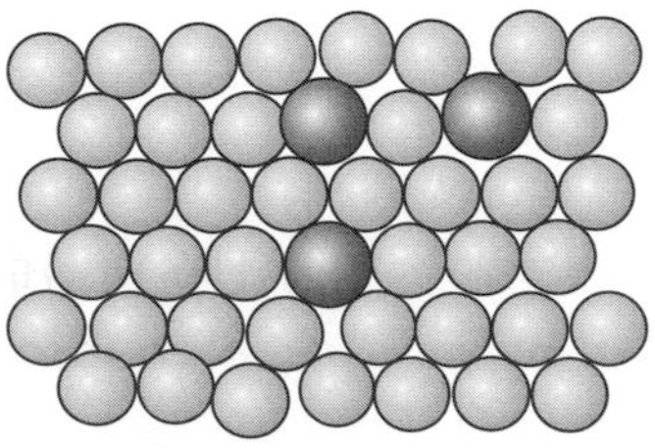

(a) What is an **alloy**?

.... *(1 mark)*

(b) What **one** word describes this general structure of metals and alloys? ....
(1 mark)

(c) By referring to the above diagram, explain why an alloy is often harder than the pure metal.

....

.... *(2 marks)*

Had a go ☐ Nearly there ☐ Nailed it! ☐

Polymers

D-C **1** Polymers have replaced many natural, traditional materials because of their cost and useful properties. The most common properties of polymers include their strength, low density and flexibility, and that they are insulators of heat and electricity.

Guided

(a) Describe the general structure of a polymer.

Polymers consist of long ....

.... *(2 marks)*

(b) For each of the examples given below, suggest **one** possible reason why the polymer is a better option than the named traditional material.

Use	Traditional material	Replacement polymer	Reason for replacement
down pipes and gutters	iron	PVC	
food bags	paper	poly(ethene)	

(2 marks)

B-A* **2** Polymers can be classified as thermosoftening or thermosetting. The diagrams below show the main differences in structure between these polymers.

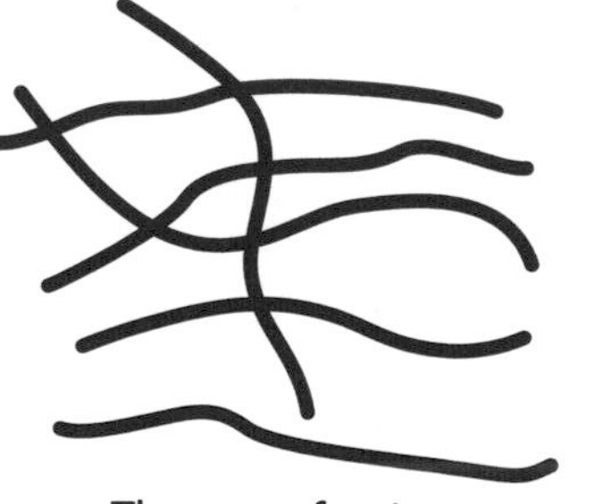

Thermosoftening

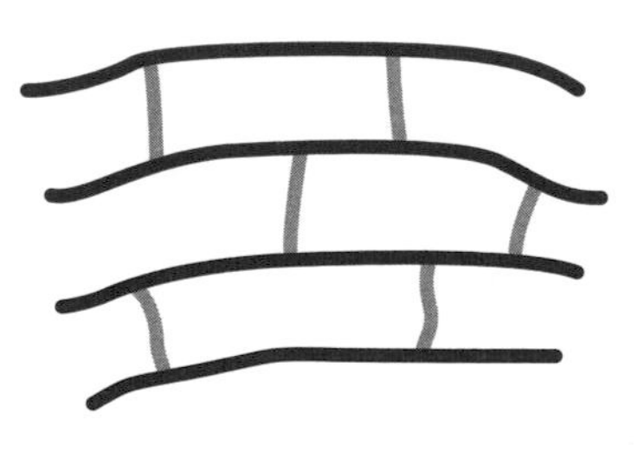

Thermosetting

(a) Describe the difference in properties between thermosoftening and thermosetting polymers.

....

.... *(2 marks)*

(b) Should the handles of cooking pots be made from a thermosoftening or a thermosetting polymer? Briefly explain your answer.

....

.... *(2 marks)*

(c) Use the diagrams to help you explain how the difference in structure between thermosoftening and thermosetting polymers gives rise to different properties.

....

....

....

.... *(4 marks)*

Remember that when you are asked to explain something, the reasons must be linked with the facts.

Nanoscience

D-C **1** Nanoscience is a branch of research into the properties of nanoparticles. Nanoparticles have found applications in a number of different areas.

(a) What are **nanoparticles**?

....

.... *(1 mark)*

Guided **(b)** Describe how the surface area of nanoparticles compares with that of ordinary powder particles.

The surface area of nanoparticles is very .... compared with the particles in ordinary powders. *(1 mark)*

D-C **2** Read the information in the box and then answer the questions.

> **Nanotechnology, good or bad?**
> Nanoparticles have been developed with many interesting new properties, and these are used in sunscreens, drug delivery, catalysts and computing. However, some scientists are concerned about the introduction of new nanotechnology. As they are so small, nanoparticles can get everywhere and can be absorbed into any part of the human body. Their full effects are unpredictable as we don't yet know all their properties. For example, silver nanoparticles, which can be used in place of antibiotics to kill bacteria, might damage other cells in different parts of the body. We therefore have to do more research, and control the trials and use of this new technology very carefully.

(a) Why are nanoparticles potentially so useful?

.... *(1 mark)*

(b) What general property on nanoparticles would make them useful as catalysts?

.... *(1 mark)*

(c) Why do we have to control the introduction of nanotechnology?

....

.... *(2 marks)*

B-A* **3** Some elements and compounds are listed in the box below.

> **carbon dioxide** **lithium** **magnesium oxide** **oxygen** **poly(ethene)** **potassium chloride**

Which of these substances fit the following descriptions?

(a) Contain delocalised electrons: .... *(1 mark)*

(b) Very large molecules but not a lattice structure: .... *(1 mark)*

Chemistry six mark question 1

The properties of a substance depend on its bonding and structure.

Explain the structure and properties of sodium chloride and hydrogen chloride in terms of the bonding found within the two substances. You should explain why sodium chloride is a solid at room temperature while hydrogen chloride is a gas. You should include labelled diagrams in your answer.

You will be more successful in six mark questions if you plan your answer before you start writing.

Your answer should include the following:

- the types of bonding and structure formed
- labelled diagrams of structures
- a description of the forces of attraction involved
- an explanation of the different melting points.

(6 marks)

Atomic structure and isotopes

1 **(a)** Complete the table with information about the three subatomic particles.

Name of particle	Charge	Relative mass	Where found in atom
proton		1	
neutron			in the nucleus
electron	negative		

(3 marks)

(b) An atom has an atomic number of 21 and a mass number of 45. What does this tell you about the structure of the atom?

.... *(3 marks)*

2 Information about the particles in two different atoms of iron is shown here: $^{56}_{26}Fe$ and $^{58}_{26}Fe$

(a) What is different about the atomic structures of these two iron atoms?

.... *(1 mark)*

(b) Show an atom of iodine with 53 protons, 74 neutrons and 53 electrons in the same way.

.... *(2 marks)*

D-C

3 Complete the missing information in the table below

Atom	Atomic number	Mass number	Number of protons	Number of neutrons	Number of electrons
$^{59}_{28}Ni$					
			38	52	38
	9			10	9

(3 marks)

Remember to check you have not mixed up mass number and atomic number. The mass number is usually the biggest.

4 Atom X has a mass number of 207 and an atomic number of 82. Atom Y has an mass number of 207 and an atomic number of 83.

(a) How many protons, neutrons and electrons are in atoms X and Y?

X = ... protons .... neutrons .... electrons

Y = ... protons .... neutrons .... electrons *(2 marks)*

(b) Explain whether atoms X and Y are atoms of the same element.

....

.... *(2 marks)*

5 Neon gas contains two types of atom, $^{22}_{10}Ne$ and $^{20}_{10}Ne$. The relative atomic mass (A_r) of neon is 20.2.

(a) What term can be used to describe these atoms? .... *(1 mark)*

(b) What is the relative atomic mass of an element? ....

.... *(2 marks)*

(c) What does the relative atomic mass of neon tell you about the abundance of $^{22}_{10}Ne$ and $^{20}_{10}Ne$?

.... *(1 mark)*

Had a go ☐ Nearly there ☐ Nailed it! ☐

Relative formula mass

Use the relative atomic masses on the periodic table on page 118 to help you answer these questions.

D-C **1** Calculate the relative formula mass (M_r) for the following substances:

Guided

(a) Sodium oxide, Na_2O.

Na_2O

→ $1 \times 16 = 16$

→ $2 \times 23 =$ ....

$=$ .... *(2 marks)*

(b) Sucrose, molecular formula $= C_{12}H_{22}O_{11}$.

....

....

....

....

$M_r =$ .... *(2 marks)*

(c) Ethyl ethanoate, structural formula =

```
   H  O     H  H
   |  ||    |  |
H—C—C—O—C—C—H
   |        |  |
   H        H  H
```

....

....

....

....

$M_r =$ .... *(2 marks)*

D-C **2** Calculate the mass of the following.

(a) 1 mole of phosphorous hydride, PH_3.

....

....

1 mole = ... g *(2 marks)*

(b) 2 moles of sulfuric acid, H_2SO_4.

....

....

2 moles = ... g *(2 marks)*

B-A* **3** Calculate the number of moles in the following:

(a) 88 g of carbon dioxide, CO_2.

....

....

Number of moles = ... *(2 marks)*

(b) 4 g of oxygen, O_2.

....

....

Number of moles = ... *(2 marks)*

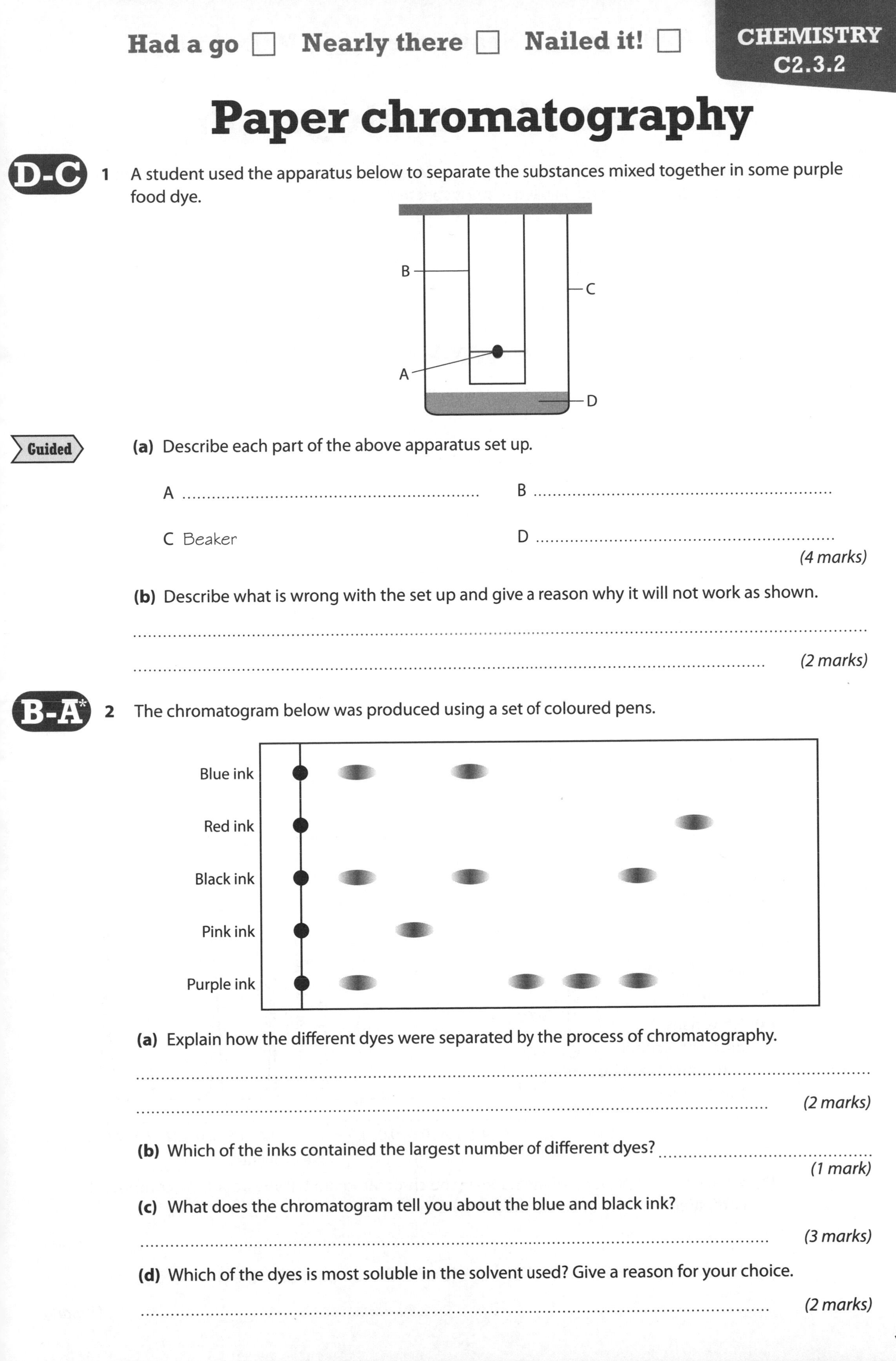

Paper chromatography

D-C

1 A student used the apparatus below to separate the substances mixed together in some purple food dye.

Guided

(a) Describe each part of the above apparatus set up.

A .... B ....

C Beaker D ....

(4 marks)

(b) Describe what is wrong with the set up and give a reason why it will not work as shown.

....

.... *(2 marks)*

B-A*

2 The chromatogram below was produced using a set of coloured pens.

(a) Explain how the different dyes were separated by the process of chromatography.

....

.... *(2 marks)*

(b) Which of the inks contained the largest number of different dyes? .... *(1 mark)*

(c) What does the chromatogram tell you about the blue and black ink?

.... *(3 marks)*

(d) Which of the dyes is most soluble in the solvent used? Give a reason for your choice.

.... *(2 marks)*

Had a go ☐ Nearly there ☐ Nailed it! ☐

Gas chromatography

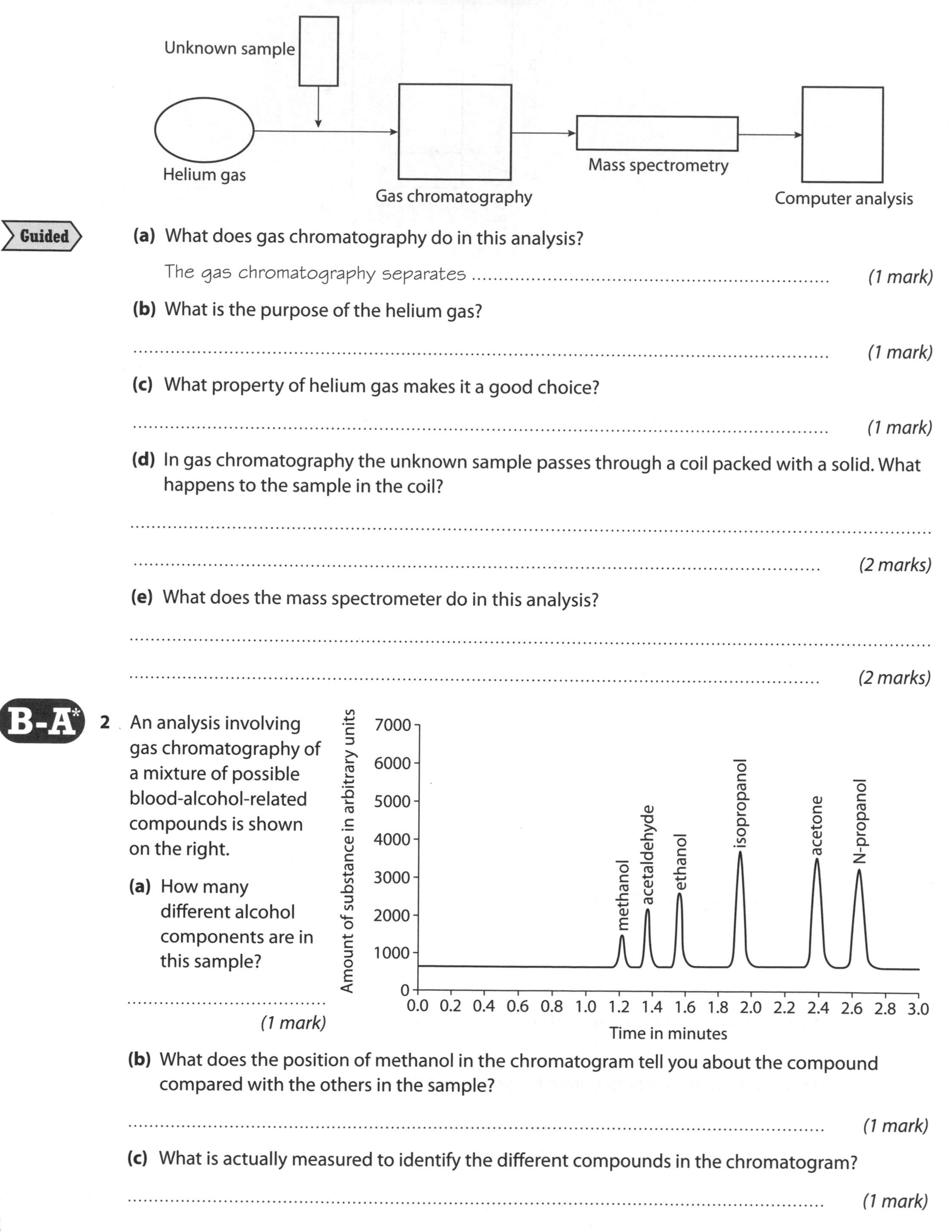

1 The flow diagram below outlines the process involved in the instrumental analysis technique called gas chromatography linked to mass spectrometry (GC-MS).

Guided

(a) What does gas chromatography do in this analysis?

The gas chromatography separates .... *(1 mark)*

(b) What is the purpose of the helium gas?

.... *(1 mark)*

(c) What property of helium gas makes it a good choice?

.... *(1 mark)*

(d) In gas chromatography the unknown sample passes through a coil packed with a solid. What happens to the sample in the coil?

....

.... *(2 marks)*

(e) What does the mass spectrometer do in this analysis?

....

.... *(2 marks)*

B-A*

2 An analysis involving gas chromatography of a mixture of possible blood-alcohol-related compounds is shown on the right.

(a) How many different alcohol components are in this sample?

.... *(1 mark)*

(b) What does the position of methanol in the chromatogram tell you about the compound compared with the others in the sample?

.... *(1 mark)*

(c) What is actually measured to identify the different compounds in the chromatogram?

.... *(1 mark)*

Chemical calculations

D-C

1 Vitamin C is used as a food supplement. Its formula is $C_6H_8O_6$. The relative formula mass (M_r) for vitamin C is 176. Calculate the percentage by mass of the following elements in vitamin C.

Guided

(a) carbon

Mass of carbon in $C_6H_8O_6$ = 6 × .... = 72

% Carbon in vitamin C = $\frac{72}{176}$ × 100% = ... %

Percentage of carbon = ... % *(2 marks)*

(b) hydrogen

....

....

Percentage of hydrogen = ... % *(2 marks)*

D-C

2 Fertilisers are made up of different compounds that contain elements which are essential to plant growth. One of these elements is nitrogen. Calculate the percentage by mass of nitrogen in each of the following compounds.

(a) potassium nitrate, KNO_3

....

....

Percentage of nitrogen = ... % *(2 marks)*

(b) ammonium chloride, NH_4Cl

....

....

Percentage of nitrogen = ... % *(2 marks)*

B-A*

3 **(a)** What is the empirical formula of a compound that was found to contain 4.5 g of carbon combined with 1.5 g of hydrogen?

....

....

.... *(3 marks)*

(b) The analysis of an 8 g sample of sulfur oxide showed that it contained 3.2 g of sulfur. Calculate the empirical formula of this sulfur oxide.

....

....

.... *(3 marks)*

B-A*

4 An organic compound X was found to contain 40% carbon, 6.7% hydrogen and 53.3% oxygen. Calculate the empirical formula of compound X.

....

....

.... *(3 marks)*

Had a go ☐ Nearly there ☐ Nailed it! ☐

Reacting mass calculations

In each of the following questions show how you work out your answer.

B-A* Guided

1 (a) The balanced equation for the combustion of methane, CH_4, is shown below.

$$CH_4 + 2O_2 \rightarrow CO_2 + 2H_2O$$

Calculate the mass of carbon dioxide produced by burning 1 g of methane. The relative formula mass of methane is 16. The relative formula mass of carbon dioxide is 44.

From the equation: 1 mole of CH_4 → 1 mole CO_2

Given the formula masses then 16 g CH_4 → .... g CO_2

Therefore 1 g CH_4 → 44/.......................... g CO_2

Mass of carbon dioxide = ... g *(3 marks)*

(b) Lime (calcium oxide) can be converted to slaked lime (calcium hydroxide) by reaction with water:

$$CaO(s) + H_2O(l) \rightarrow Ca(OH)_2(aq)$$

What mass of slaked lime could be produced from 5000 kg of lime? The relative formula mass of calcium oxide is 56. The relative formula mass of calcium hydroxide is 74.

....

....

....

Mass of slaked lime = ... kg *(3 marks)*

B-A*

2 The balanced equation for the reduction of iron(III) oxide is shown below.

$$2Fe_2O_3 + 3C \rightarrow 4Fe + 3CO_2$$

What mass of iron can be extracted by the reduction of 2000 g of iron(III) oxide? The relative formula mass of iron(III) oxide is 160. The relative atomic mass of iron is 56.

....

....

....

Mass of iron = ... g *(3 marks)*

B-A*

3 The neutralisation reaction between sodium hydroxide and sulfuric acid is represented by:

$$H_2SO_4 + 2NaOH \rightarrow Na_2SO_4 + 2H_2O$$

What mass of sodium hydroxide, NaOH, is needed to make 25 g of sodium sulfate, Na_2SO_4? The relative formula mass of sodium hydroxide is 40. The relative formula mass of sodium sulfate is 142.

....

....

....

Mass of sodium hydroxide = ... g *(3 marks)*

Reaction yields

D-C

1 During all chemical reactions reactants change into products. The amount of product obtained is called the yield. When the actual yield of product is calculated as a percentage of the calculated yield (the amount that should have been formed) it is called a percentage yield.

(a) Explain, in terms of the atoms and molecules involved, why the total mass of products should be the same as the total mass of reactants.

....

.... *(2 marks)*

(b) Describe **three** possible reasons for the actual yield being less than the expected yield in a particular chemical change.

....

....

.... *(3 marks)*

Guided

(c) Use the information in the passage to complete the equation below, to show how the percentage yield in a chemical reaction could be calculated.

$$\% \text{ yield} = \frac{\qquad}{\text{calculated yield}} \times \text{.............................}$$

(1 mark)

B-A*

2 A student was investigating the formation of magnesium oxide by burning magnesium in air.

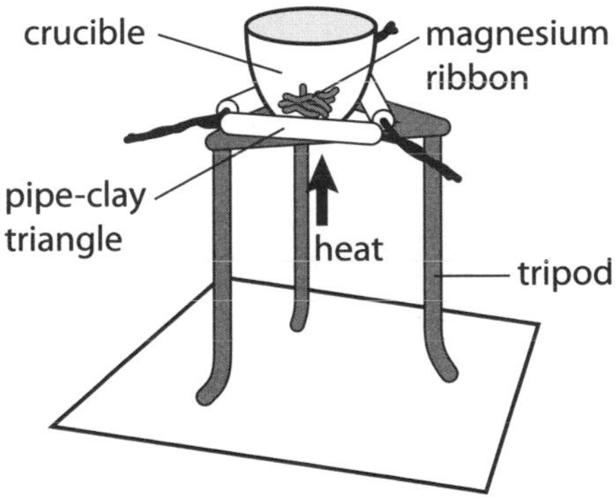

The balanced equation for the reaction is:

$$2Mg + O_2 \rightarrow 2MgO.$$

The student calculated that 1.2 g of magnesium should react to produce 2.0 g of magnesium oxide.

(a) What mass of oxygen would combine with 1.2 g of magnesium to produce 2.0 g of magnesium oxide?

.... *(1 mark)*

(b) If the experiment had a 50% yield how much magnesium oxide would be obtained?

.... *(1 mark)*

(c) Suggest the **two** most likely reasons for the low yield of magnesium oxide. In each case explain your reason with reference to the experimental set-up shown.

....

....

....

.... *(4 marks)*

Had a go ☐ Nearly there ☐ Nailed it! ☐

Reversible reactions

D-C

1 Methane can be formed when carbon monoxide reacts with hydrogen.

$$CO(g) + 3H_2(g) \rightleftharpoons CH_4(g) + H_2O(g)$$

Guided

(a) What does the double arrow ($\rightleftharpoons$) between reactants and products mean?

This means the reaction goes .... *(1 mark)*

AQA SKILL Interpret Page 121

(b) Name the molecules that will be present when this reaction has been left for some time.

.... *(2 marks)*

B-A*

2 In the reaction shown in question **1**, calculations show that 28 g of carbon monoxide should produce 16 g of methane. If the percentage yield is 20%, what will be the actual mass of methane obtained?

AQA SKILL Interpret Page 121

....

....

Mass of methane = ... g *(2 marks)*

B-A*

3 Ammonia (NH_3) gas is manufactured by the Haber process. This uses nitrogen (N_2) from the air, and hydrogen (H_2) made from methane and water. The reaction is reversible and the reaction in the ammonia converter only produces a 5% yield of ammonia.

AQA SKILL Explain Page 121

$$N_2(g) + 3H_2(g) \rightleftharpoons 2NH_3(g)$$

A flow diagram for the Haber process is shown in the diagram.

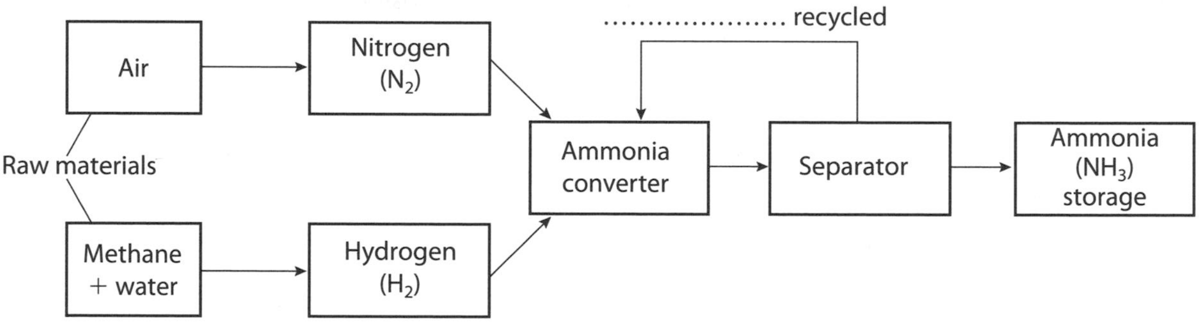

(a) What is recycled and how does this make the process sustainable by reducing costs and increasing profit?

....

.... *(2 marks)*

(b) (i) Using the balanced equation given above, calculate the mass of ammonia that could theoretically be produced from 72 g of nitrogen gas. The relative formula mass of ammonia is 17. The relative formula mass of nitrogen is 28.

....

....

Mass of ammonia = ... g *(3 marks)*

(ii) Calculate the actual mass of ammonia which is produced if the yield is 5%.

....

....

Mass of ammonia = ... g *(2 marks)*

Rates of reaction 1

D-C

1 A student carried out an experiment to investigate the rate of reaction between marble chips and hydrochloric acid. To follow the reaction rate the student measured the mass lost by the reaction mixture with time. The results of the experiment are shown below.

Time in min	0	1	2	3	4	5	6	7	8	9	10
Mass lost in g	0	0.12	0.22	0.30	0.36	0.40	0.42	0.45	0.45	0.45	0.45

(a) Calculate the rate in g/min between:

Guided

(i) 2 and 4 seconds Rate = $\frac{\text{change}}{\text{time}} = \frac{0.36 - 0.22}{4 - 2}$

= .... Rate = ... g/min *(2 marks)*

(ii) 4 and 6 seconds ....

....

Rate = ... g/min *(2 marks)*

(b) Explain when this reaction is completed.

....

.... *(2 marks)*

B-A*

2 The rate of reaction between calcium carbonate and hydrochloric acid was investigated by using calcium carbonate lumps and powder.

$$2HCl(aq) + CaCO_3(s) \rightarrow CuCl_2(aq) + H_2O(l) + CO_2(g)$$

The graph on the right shows the results of measuring the volume of gas produced against time for two different experiments.

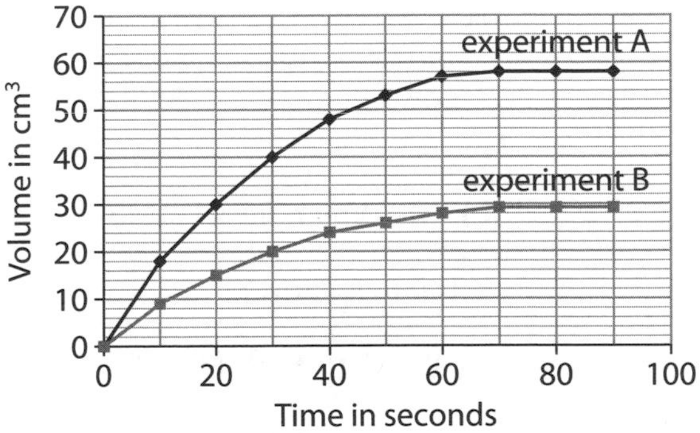

(a) Calculate the mean rate of reaction in cm^3/s between 0 and 20 seconds in experiment A.

....

....

....

Rate of reaction = ... cm^3/s *(2 marks)*

(b) Calculate the rate of reaction in cm^3/s between 0 and 20 seconds in experiment B.

....

....

Rate of reaction = ... cm^3/s *(2 marks)*

(c) At what time are both reactions finished? .... *(1 mark)*

(d) Did both experiments, A and B, use the same amount of reactants? Explain your answer.

....

....

.... *(3 marks)*

Rates of reaction 2

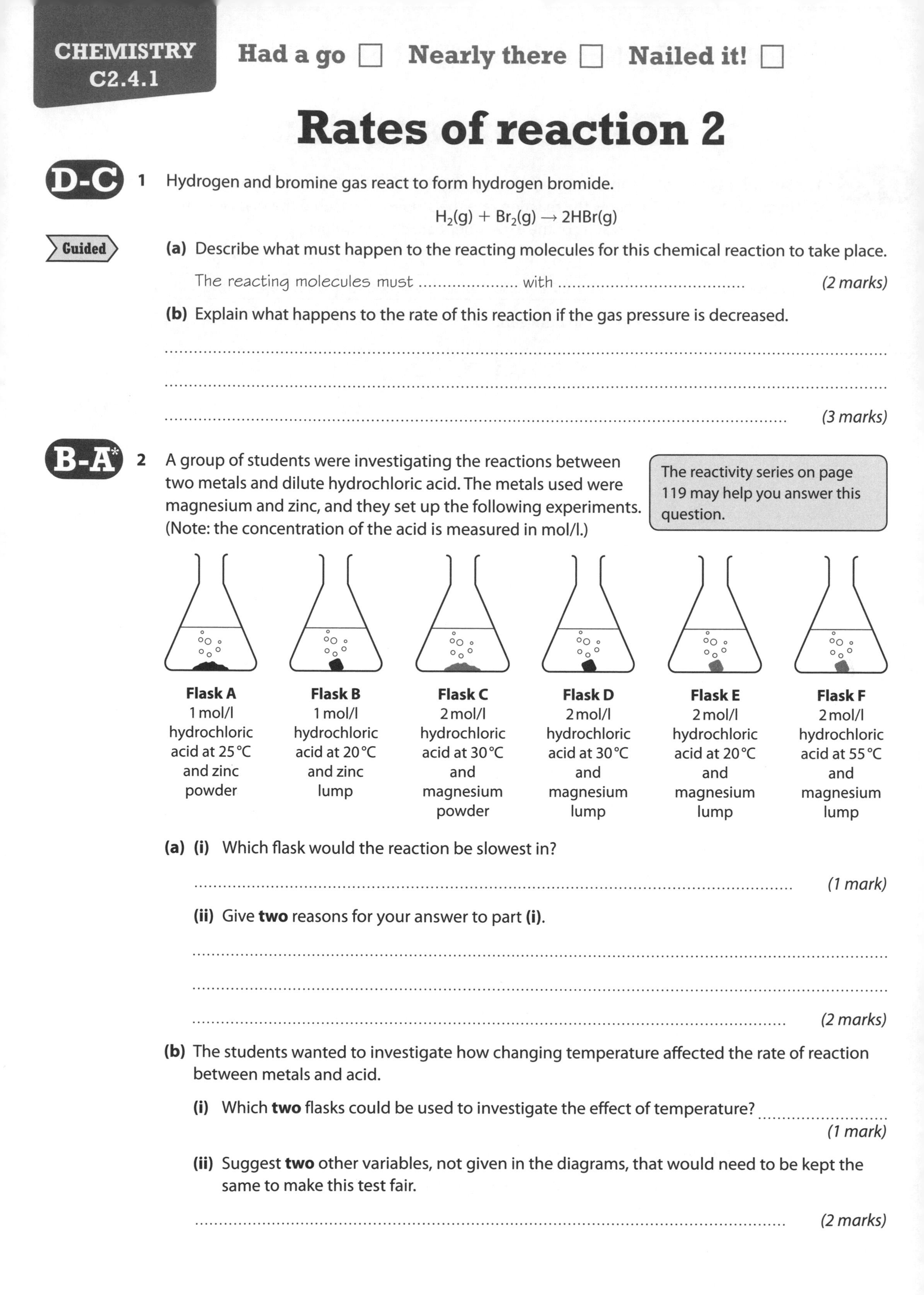

D-C **1** Hydrogen and bromine gas react to form hydrogen bromide.

$$H_2(g) + Br_2(g) \rightarrow 2HBr(g)$$

Guided **(a)** Describe what must happen to the reacting molecules for this chemical reaction to take place.

The reacting molecules must .... with .... *(2 marks)*

(b) Explain what happens to the rate of this reaction if the gas pressure is decreased.

....

....

.... *(3 marks)*

B-A* **2** A group of students were investigating the reactions between two metals and dilute hydrochloric acid. The metals used were magnesium and zinc, and they set up the following experiments. (Note: the concentration of the acid is measured in mol/l.)

The reactivity series on page 119 may help you answer this question.

Flask A 1 mol/l hydrochloric acid at 25 °C and zinc powder

Flask B 1 mol/l hydrochloric acid at 20 °C and zinc lump

Flask C 2 mol/l hydrochloric acid at 30 °C and magnesium powder

Flask D 2 mol/l hydrochloric acid at 30 °C and magnesium lump

Flask E 2 mol/l hydrochloric acid at 20 °C and magnesium lump

Flask F 2 mol/l hydrochloric acid at 55 °C and magnesium lump

(a) (i) Which flask would the reaction be slowest in?

.... *(1 mark)*

(ii) Give **two** reasons for your answer to part **(i)**.

....

....

.... *(2 marks)*

(b) The students wanted to investigate how changing temperature affected the rate of reaction between metals and acid.

(i) Which **two** flasks could be used to investigate the effect of temperature? .... *(1 mark)*

(ii) Suggest **two** other variables, not given in the diagrams, that would need to be kept the same to make this test fair.

.... *(2 marks)*

Chemistry six mark question 2

The yields in reactions in the chemical industry are often less than 100%. This is a problem that has to be overcome when running an industrial chemical process.

Describe the meaning of a low yield and why it is a problem in industrial processes. Describe two reasons why a reaction might produce less than a 100% yield, and suggest at least one way that the problem of a low yield can be overcome.

You will be more successful in six mark questions if you plan your answer before you start writing.
Your answer should include the following:

- the meaning of yield, and the problem of a low yield in an industrial process
- a description of at least two reasons for a low yield
- a description of changes that can be made to improve the yield in a chemical process.

(6 marks)

Energy changes

D-C **1** Chemical energy changes have many uses in industry and the home. These changes, which can be exothermic or endothermic, all involve a transfer of heat energy from one place to another.

Guided **(a)** Describe the difference between exothermic and endothermic reactions.

An exothermic reaction .... heat energy while an endothermic reaction .... heat energy. *(2 marks)*

(b) How could you tell if an endothermic reaction was taking place in a solution?

....

.... *(2 marks)*

(c) Give **one** example of a chemical reaction that is exothermic.

.... *(1 mark)*

B-A* **2** A group of students were given the apparatus listed below and asked to investigate the neutralisation reactions between two different acids and an alkali. They were asked to find out which neutralisation reaction was most exothermic.

0.5 mol/l hydrochloric acid (acid)	2 × thermometers
0.5 mol/l nitric acid (acid)	2 × measuring cylinders
0.5 mol/l sodium hydroxide (alkali)	2 × insulated cups

(a) Describe the method the students would use for this investigation by writing down the steps they would carry out. The first step has been written for you.

A: Measure 25 cm^3 of 0.5 mol/l hydrochloric acid, in measuring cylinder, and put into insulated cup.

....

....

....

.... *(3 marks)*

(b) Describe **two** things the students would need to do to make the test fair.

....

.... *(2 marks)*

(c) How could the students use their results to tell which neutralisation was most exothermic?

.... *(1 mark)*

B-A* **3** The equation for the combustion of glucose is shown below.

$$C_6H_{12}O_6 + 6O_2 \rightarrow 6CO_2 + 6H_2O \text{ (+ energy)}$$

The reaction in plants that forms glucose, called photosynthesis, is shown below.

$$6CO_2 + 6H_2O \rightarrow C_6H_{12}O_6 + 6O_2$$

Use the information given above to explain if photosynthesis is an exothermic or endothermic reaction.

....

.... *(2 marks)*

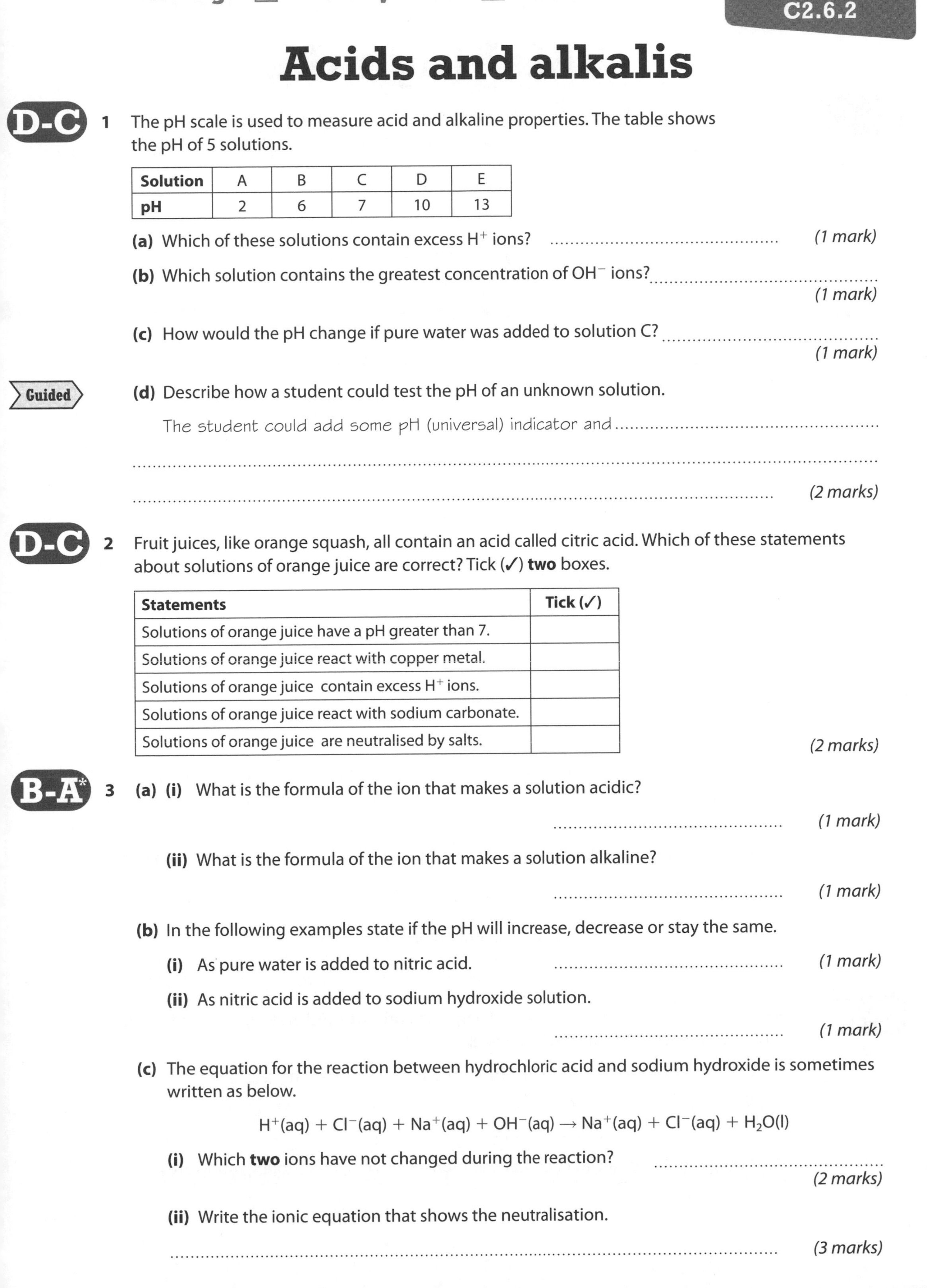

Acids and alkalis

D-C

1 The pH scale is used to measure acid and alkaline properties. The table shows the pH of 5 solutions.

Solution	A	B	C	D	E
pH	2	6	7	10	13

(a) Which of these solutions contain excess H^+ ions? .... *(1 mark)*

(b) Which solution contains the greatest concentration of OH^- ions? .... *(1 mark)*

(c) How would the pH change if pure water was added to solution C? .... *(1 mark)*

Guided

(d) Describe how a student could test the pH of an unknown solution.

The student could add some pH (universal) indicator and ....

....

.... *(2 marks)*

D-C

2 Fruit juices, like orange squash, all contain an acid called citric acid. Which of these statements about solutions of orange juice are correct? Tick (✓) **two** boxes.

Statements	Tick (✓)
Solutions of orange juice have a pH greater than 7.	
Solutions of orange juice react with copper metal.	
Solutions of orange juice contain excess H^+ ions.	
Solutions of orange juice react with sodium carbonate.	
Solutions of orange juice are neutralised by salts.	

(2 marks)

B-A*

3 (a) (i) What is the formula of the ion that makes a solution acidic?

.... *(1 mark)*

(ii) What is the formula of the ion that makes a solution alkaline?

.... *(1 mark)*

(b) In the following examples state if the pH will increase, decrease or stay the same.

(i) As pure water is added to nitric acid. .... *(1 mark)*

(ii) As nitric acid is added to sodium hydroxide solution.

.... *(1 mark)*

(c) The equation for the reaction between hydrochloric acid and sodium hydroxide is sometimes written as below.

$$H^+(aq) + Cl^-(aq) + Na^+(aq) + OH^-(aq) \rightarrow Na^+(aq) + Cl^-(aq) + H_2O(l)$$

(i) Which **two** ions have not changed during the reaction? .... *(2 marks)*

(ii) Write the ionic equation that shows the neutralisation.

.... *(3 marks)*

Had a go ☐ Nearly there ☐ Nailed it! ☐

Making salts

D-C 1 **(a)** What **two** products are formed during all neutralisation reactions?

Guided

Salt and .... *(2 marks)*

(b) Which **two** ions react during all neutralisation reactions?

.... *(2 marks)*

(c) How does the pH of acids change when they are neutralised?

.... *(1 mark)*

D-C 2 Indigestion is caused by too much hydrochloric acid in the stomach. Some indigestion remedies contain the insoluble compounds magnesium hydroxide and aluminium hydroxide to react with the excess hydrochloric acid.

(a) What one-word term can be used to describe magnesium hydroxide and aluminium hydroxide?

....

(1 mark)

(b) Name the **two** salts formed when magnesium hydroxide and aluminium hydroxide react with the excess hydrochloric acid.

.... *(2 marks)*

D-C 3 **(a)** Complete this table, which shows the acids and bases that can be used to make certain salts.

Acid	Base	Salt
hydrochloric acid	lithium hydroxide	
	calcium oxide	calcium nitrate
sulfuric acid		iron(II) sulfate

(3 marks)

(b) Ammonium hydroxide, NH_4OH, reacts with dilute sulfuric acid, H_2SO_4, to form the soluble salt ammonium sulfate, $(NH_4)_2SO_4$, and water, H_2O. Ammonium sulfate is a very valuable industrial chemical.

(i) Complete the balanced equation, with state symbols, for this neutralisation reaction.

2 ....(......) + ...(aq) → ....(......) + 2 ....(l) *(2 marks)*

(ii) Suggest a possible use for ammonium sulfate.

.... *(1 mark)*

B-A* 4 A student wished to find the volume of a hydrochloric acid, HCl, needed to exactly neutralise 20 cm^3 of a solution of calcium hydroxide, $Ca(OH_2)$.

(a) Explain what the student should do to find the point when the calcium hydroxide was exactly neutralised.

.... *(2 marks)*

(b) Complete the balanced chemical equation for this reaction (state symbols are not required)

.... + ... → .... + $2H_2O$ *(3 marks)*

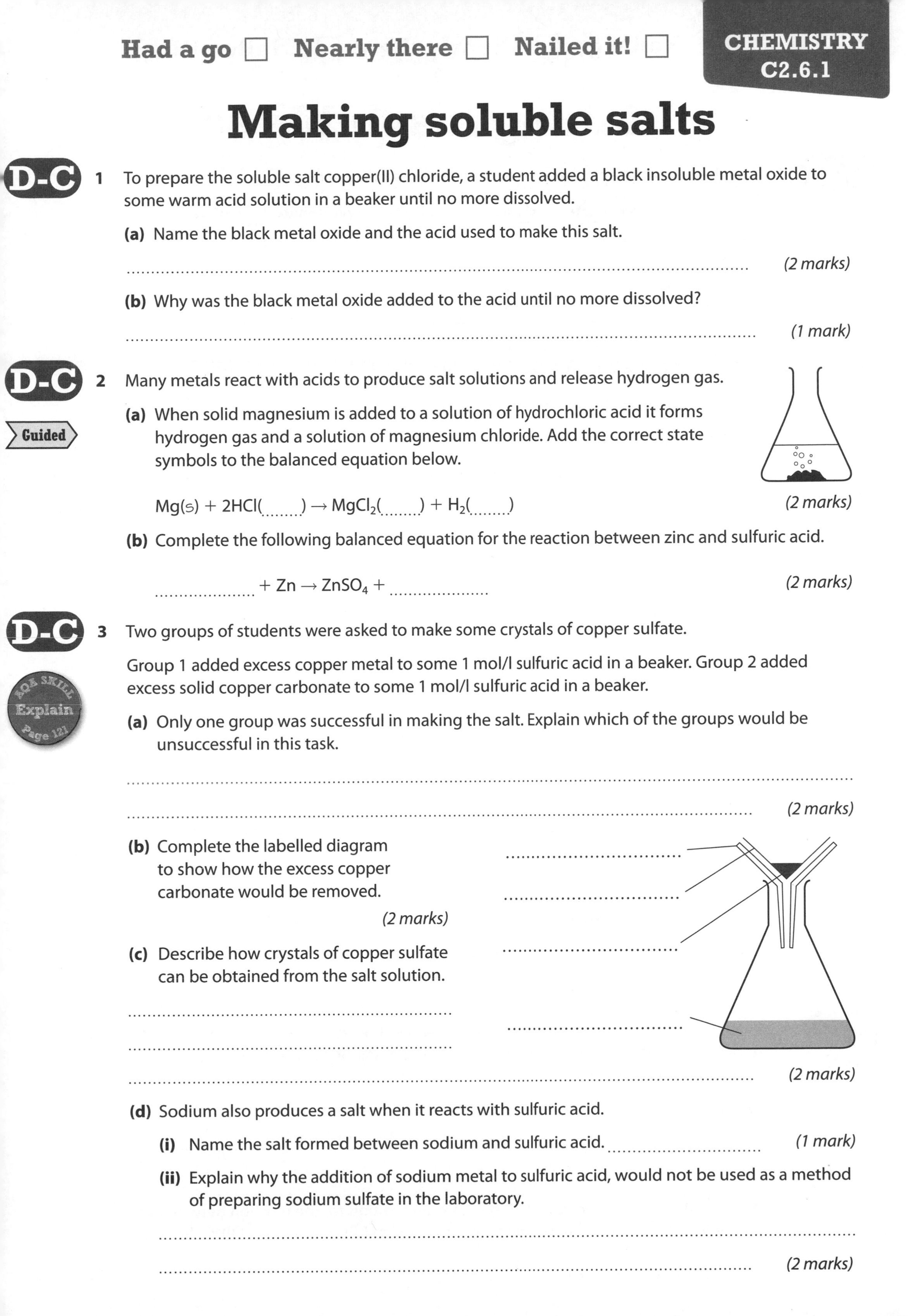

Making soluble salts

D-C **1** To prepare the soluble salt copper(II) chloride, a student added a black insoluble metal oxide to some warm acid solution in a beaker until no more dissolved.

(a) Name the black metal oxide and the acid used to make this salt.

.... *(2 marks)*

(b) Why was the black metal oxide added to the acid until no more dissolved?

.... *(1 mark)*

D-C **2** Many metals react with acids to produce salt solutions and release hydrogen gas.

Guided **(a)** When solid magnesium is added to a solution of hydrochloric acid it forms hydrogen gas and a solution of magnesium chloride. Add the correct state symbols to the balanced equation below.

$Mg(s) + 2HCl(\ldots\ldots) \rightarrow MgCl_2(\ldots\ldots) + H_2(\ldots\ldots)$ *(2 marks)*

(b) Complete the following balanced equation for the reaction between zinc and sulfuric acid.

$\ldots\ldots\ldots\ldots + Zn \rightarrow ZnSO_4 + \ldots\ldots\ldots\ldots$ *(2 marks)*

D-C **3** Two groups of students were asked to make some crystals of copper sulfate.

AQA SKILL Explain Page 121

Group 1 added excess copper metal to some 1 mol/l sulfuric acid in a beaker. Group 2 added excess solid copper carbonate to some 1 mol/l sulfuric acid in a beaker.

(a) Only one group was successful in making the salt. Explain which of the groups would be unsuccessful in this task.

....

.... *(2 marks)*

(b) Complete the labelled diagram to show how the excess copper carbonate would be removed. *(2 marks)*

....

....

....

....

(c) Describe how crystals of copper sulfate can be obtained from the salt solution.

....

....

.... *(2 marks)*

(d) Sodium also produces a salt when it reacts with sulfuric acid.

(i) Name the salt formed between sodium and sulfuric acid. .... *(1 mark)*

(ii) Explain why the addition of sodium metal to sulfuric acid, would not be used as a method of preparing sodium sulfate in the laboratory.

....

.... *(2 marks)*

Had a go ☐ Nearly there ☐ Nailed it! ☐

Making insoluble salts

D-C **1** Use the solubility table below to help you to answer these questions.

	carbonate	chloride	nitrate	phosphate	sulfate
calcium	i	s	s	i	ss
copper(II)	i	s	s	i	s
magnesium	i	s	s	i	s
lead(II)	i	ss	s	i	i
sodium	s	s	s	s	s
nickel	i	s	s	i	s

Key: s = soluble; ss = slightly soluble; i = insoluble

(a) Name the most soluble salt of lead in the table. .... *(1 mark)*

(b) Name **two** solutions that could be used to make solid samples of the following insoluble salts.

Guided **(i)** lead(II) phosphate: lead nitrate and sodium phosphate *(2 marks)*

(ii) copper(II) carbonate: .... and .... *(2 marks)*

(c) What process would be used to separate the insoluble salt from the solution?

.... *(1 mark)*

B-A* **2** A student carried out the following experiment.

(a) Name a different metal carbonate that could be used to make the same insoluble salt.

.... *(1 mark)*

(b) Complete the balanced equation including state symbols.

$K_2CO_3(aq) + 2AgNO_3(aq) \rightarrow Ag_2CO_3$ (......) + ... (aq) *(3 marks)*

(c) State why this is **not** a suitable method for making sodium sulfate.

.... *(1 mark)*

B-A* **3** Precipitation reactions can be used in the treatment of waste water. Different reagents are used depending on the ion to be removed. For example, solutions of sodium hydroxide are commonly used to remove certain transition metal ions.

(a) What is a precipitation reaction and how does it remove ions from solution?

....

.... *(2 marks)*

(b) Complete the balanced equation with state symbols for the reaction between sodium hydroxide and nickel(II) chloride solutions, which forms solid nickel(II) hydroxide.

$2NaOH(aq) + NiCl_2(aq) \rightarrow$ ....(......) + ...(......) *(2 marks)*

Using electricity

D-C

1 Molten lead bromide breaks down when it conducts electricity.

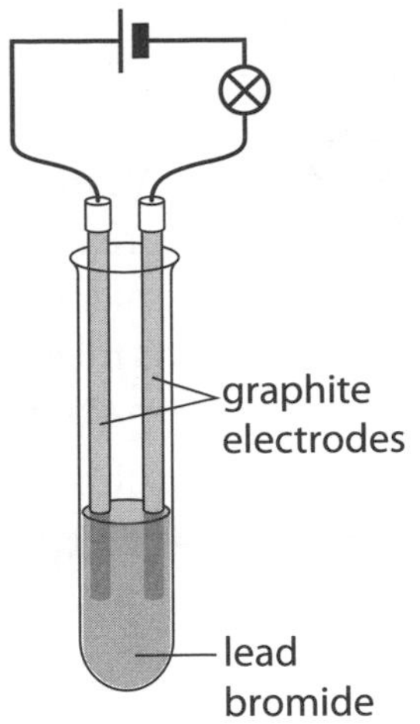

AQA SKILL Predict Page 121

(a) Using the apparatus shown, how would you know if an electric current was passing?

....

(1 mark)

graphite electrodes

lead bromide

Guided

(b) Explain why the lead bromide needs to be molten.

Solid lead bromide ....

....

(2 marks)

(c) Name the products at each of the electrodes during this process.

....

.... *(2 marks)*

(d) What is a reduction reaction and where does it occur during this process?

....

.... *(2 marks)*

D-C

2 **(a)** Explain what happens to metal ions during electrolysis. ....

.... *(2 marks)*

(b) What is the name of the process that involves loss of electrons? ....

(1 mark)

B-A*

3 The electrolysis of copper(II) chloride solution breaks up the compound, and forms copper and chlorine.

AQA SKILL Predict Page 121

Explain what happens when a direct electric current is passed through copper(II) chloride solution. Write ionic equations for the reactions at the electrodes.

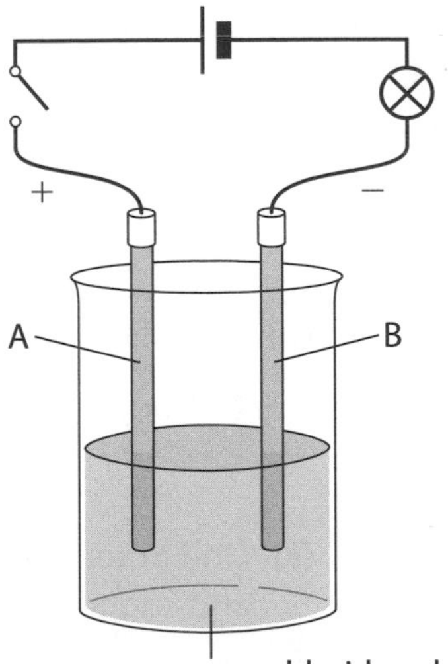

+ −

A B

copper chloride solution

....

....

....

....

....

....

....

....

....

....

.... *(6 marks)*

Useful substances from electrolysis

D-C **1** Aluminium is extracted from its ore using electricity. The ore, which mainly contains aluminium oxide, is mixed with cryolite before it is melted and electrolysed.

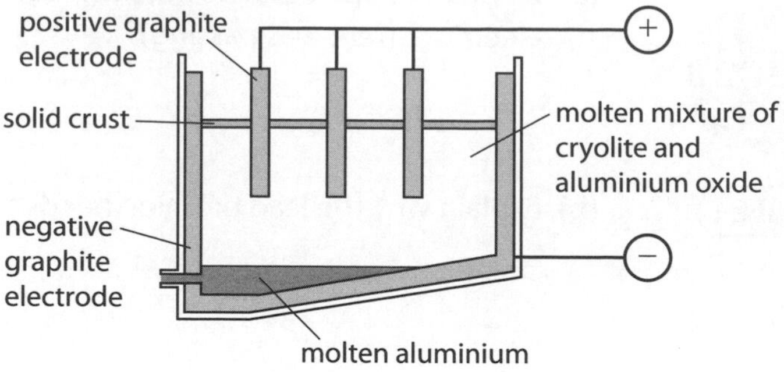

positive graphite electrode
solid crust
molten mixture of cryolite and aluminium oxide
negative graphite electrode
molten aluminium
+
−

(a) Explain why molten aluminium oxide conducts electricity while the solid does not.

....

.... *(2 marks)*

(b) Why is cryolite added to the aluminium oxide?

.... *(1 mark)*

Guided **(c)** Why is the aluminium metal formed at the negative electrode?

Aluminium ions are positive ....

.... *(2 marks)*

(d) Name **two** products that could be formed at the positive electrode during this process.

.... *(2 marks)*

B-A* **2** The electrolysis of sodium chloride solution produces hydrogen and chlorine gas at the electrodes.

AQA SKILL Predict Page 121

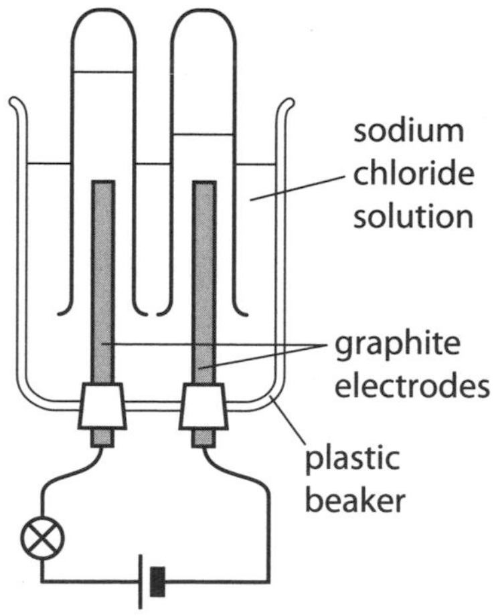

sodium chloride solution
graphite electrodes
plastic beaker

(a) At which electrode will each gas form?

....

....

.... *(2 marks)*

(b) Complete the half equations for the formation of these gases.

$2H^+(aq)$ + ... → ....

.... → $Cl_2(g)$ + ... *(4 marks)*

(c) One other important product is formed in solution during this electrolysis. Name the product and suggest a possible use for it.

....

.... *(2 marks)*

(d) Name **two** consumer products that use chlorine in their manufacture.

....

.... *(2 marks)*

Electrolysis products

AQA SKILL Predict Page 121

1 Electroplating can be used to put a coating of metals like copper and silver onto iron. Electroplating always involves the reduction of metal ions during electrolysis.

The table below shows the products formed during the electrolysis of some electrolyte solutions.

Electrolyte solution	Product at positive electrode	Product at negative electrode
sodium chloride		hydrogen
copper iodide		copper
magnesium bromide	bromine	

(a) Complete the missing information in the table. *(3 marks)*

Guided

(b) What is an electrolyte?

A solution .... as it .... *(2 marks)*

(c) What happens to metal ions during the reduction reaction?

.... *(1 mark)*

(d) Electroplating does not work with all metals.

(i) What kind of metals cannot be electroplated onto other metals?

.... *(1 mark)*

(ii) If the metal is not formed during electrolysis of a solution, what is usually produced at the negative electrode instead?

.... *(1 mark)*

D-C

2 Electroplating is used to place a thin coating of an expensive metal onto metal objects. The diagram on the right shows how to silver plate a copper ring.

power supply
positive electrode
negative electrode
silver metal
copper ring
silver ion solution

(a) Suggest a reason for electroplating objects.

....

.... *(1 mark)*

(b) Which electrode does the item to be electroplated become?

.... *(1 mark)*

(c) During the electroplating what carries the electric current:

(i) through the wires in the circuit? .... *(1 mark)*

(ii) through the solution? .... *(1 mark)*

B-A*

3 **(a)** Complete the half equation to show how the silver ion (Ag^{+}) changes into silver metal on the copper ring. Include state symbols in this equation.

$Ag^{+}(aq)$ + ... → .... *(2 marks)*

(b) Electroplating is expensive and can be difficult carry out. Dipping in molten metal is a much simpler process. Suggest a reason why electroplating is preferred for coating certain metals.

....

.... *(2 marks)*

Chemistry six mark question 3

You have been asked to make a sample of cobalt chloride, a soluble salt with a deep red colour. Describe the method you would use to make a dry sample of solid cobalt chloride.

You should use some, but not all, of the apparatus and chemicals listed below.

hydrochloric acid solution	2 × 250 cm^3 beakers
sulfuric acid solution	spatula
cobalt nitrate solution	evaporating basin
cobalt oxide solid (insoluble)	filter funnel and filter paper
2 × 250 cm^3 conical flasks	Bunsen burner and heat mat

You will be more successful in six mark questions if you plan your answer before you start writing. Remember to include the following in your answer:

- the chemicals you will use and a description of how you know the reaction is completed
- a description of how you will separate out the solid salt and any safety precautions required.

....

(6 marks)

Resultant forces

D-C **1** A box has a mass of 10 kg. The Earth exerts a force of 100 N downwards on the box.

(a) What is the name of this force?

.... *(1 mark)*

(b) State the size and direction of the force that the box exerts on the Earth.

.... *(2 marks)*

D-C **2** The diagram shows some forces on a car.

2000 N

15 000 N

4000 N

In an exam, all the equations you need will be given to you on a separate sheet. In this book, the equations sheet is on page 120. At Higher Tier you are expected to be able to rearrange equations when necessary.

Guided **(a)** Calculate the resultant force on the car.

Total forces acting backwards = 2000 N + 4000 N

= ....

Resultant = 15 000 N − ....

= ....

Resultant force = ... N forwards *(2 marks)*

Remember that the direction of a force is important, as well as its size.

(b) State how this resultant force will affect the car.

.... *(2 marks)*

B-A* **3** The diagram shows a satellite orbiting the Earth. It is travelling at 6000 m/s. The satellite is controlled using two identical small rockets called thrusters.

EXAM ALERT

When the resultant force on an object is zero don't assume that the object is stationary – it might be travelling at a constant speed.

Students have struggled with questions like this in recent exams – **be prepared!**

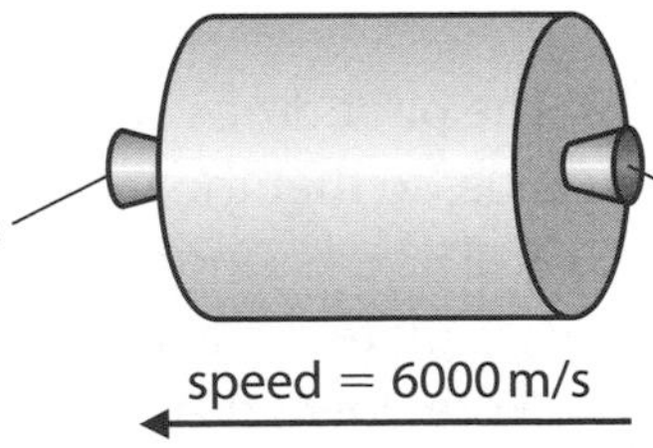

thruster A

thruster B

speed = 6000 m/s

(a) Thruster A fires for a few seconds. Explain the effect this will have on the speed of the satellite.

....

....

.... *(3 marks)*

(b) A fault develops, and thrusters A and B both fire at once. Explain what effect this will have on the speed of the satellite.

....

.... *(2 marks)*

Had a go ☐ Nearly there ☐ Nailed it! ☐

Forces and motion

D-C

1 For each of these questions, write down the equation you use, and then show clearly how you work out your answer.

Guided

(a) A lorry has a mass of 27 500 kg and accelerates at 1.5 m/s^2.
Calculate the resultant force on the lorry.

force = mass × acceleration

= 27 500 kg × 1.5 m/s^2

= ....

On every page you will find a guided question. Guided questions have part of the answer filled in for you to show you how best to answer them.

Resultant force = ... N *(2 marks)*

(b) A motorcycle has a mass of 240 kg. There is a resultant forwards force on it of 1200 N, in the direction it is moving. Calculate the acceleration.

....

....

Acceleration = ... m/s^2 *(2 marks)*

(c) A ball is dropped and accelerates at 10 m/s^2. The force on it is 0.6 N.
Calculate the mass of the ball.

....

....

Mass = ... kg *(2 marks)*

B-A*

2 A hot air balloon is moving at a constant speed of 3 m/s horizontally, but is not moving in a vertical direction. The total mass of the balloon and its passengers is 3000 kg and the weight is 30 000 N.

(a) Explain why the resultant force on the balloon is must be zero.

....

.... *(2 marks)*

(b) The pilot drops some sandbags over the side, so that the total weight of the balloon and its passengers is now 29 000 N.
Calculate the acceleration of the balloon and give its direction.
Write down the equation you use, and then show clearly how you work out your answer.

Start by calculating the resultant force on the balloon now that its weight has changed.

....

....

....

Acceleration = ... m/s^2 .... *(4 marks)*

(c) Explain how dropping the sandbags affects the horizontal speed of the balloon.

....

.... *(2 marks)*

Distance–time graphs

D-C **1** State the difference between speed and velocity.

....

.... *(2 marks)*

D-C **2** The distance–time graph shows a person walking at a constant speed.

AQA SKILL Interpret Page 121

(a) Draw a line on the graph to show a person starting from the same point and walking for the same distance at half the speed. Label this line B. *(2 marks)*

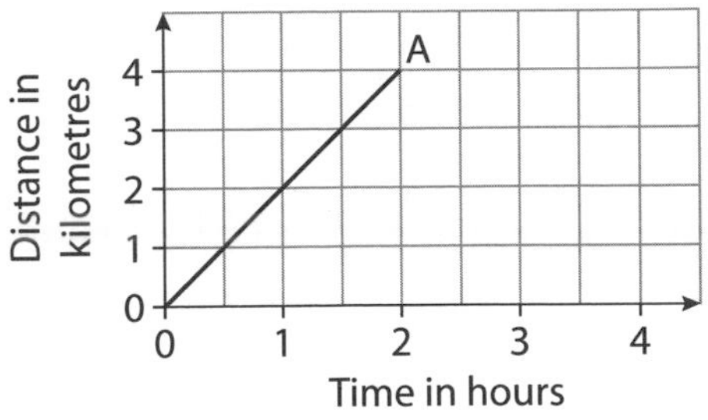

(b) Draw a line on the graph showing a person starting from the same point and running for 2 km and then stopping. Label this line C. *(2 marks)*

Guided **(c)** Calculate the speed of the walker shown by line A.

$$\text{speed} = \text{gradient} = \frac{\text{change in distance}}{\text{change in time}}$$

$$= \frac{4\text{ km}}{2\text{ hours}}$$

= ....

Speed = ... km/h *(2 marks)*

B-A* **3** The distance–time graph below shows the journey of a person to the shops and back.

AQA SKILL Interpret Page 121

(a) Calculate the speed of the person between 3 and 5 minutes.
Give your answer in metres/second.

Don't forget to convert the time to seconds before carrying out your calculation.

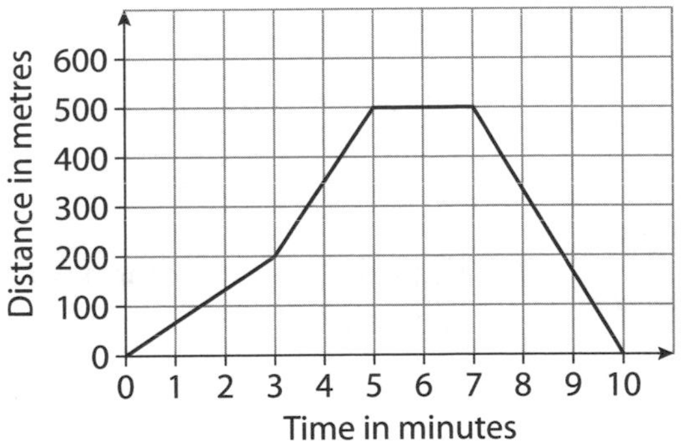

....

....

....

Speed = ... m/s *(3 marks)*

(b) Explain **two** ways in which the velocity of the person between 7 and 10 minutes is different from their velocity between 0 and 3 minutes.

....

....

....

.... *(4 marks)*

Had a go ☐ Nearly there ☐ Nailed it! ☐

Acceleration and velocity

D-C **1** Calculate the acceleration for each of the following. For each of these questions, write down the equation you use, and then show clearly how you work out your answer.

Guided

(a) A motorcycle that goes from 2 m/s to 14 m/s in 4 seconds.

$v = 14$ m/s, $u = 2$ m/s

$a = \frac{v - u}{t} = \frac{14 - \dots}{4} = \dots$

Acceleration = ... m/s² *(2 marks)*

(b) A car that changes its speed from 5 m/s to 30 m/s in 2 minutes.

Don't forget to change the units for time.

....

....

Acceleration = ... m/s² *(3 marks)*

D-C **2** Lorries are fitted with tachographs that show how fast they travel each day. The following two graphs show data for two lorries on their journeys one morning.

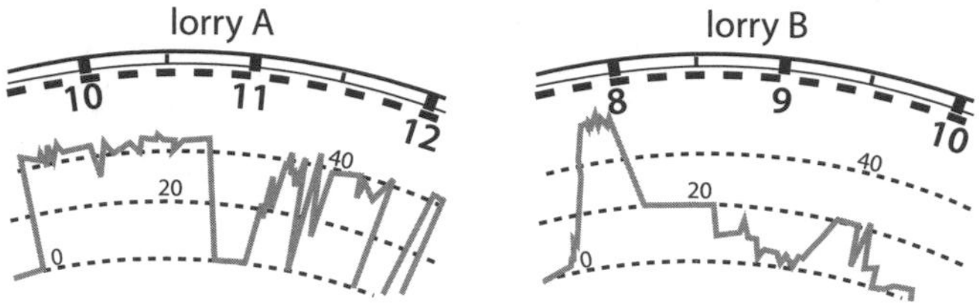

lorry A

lorry B

Use the data in the graphs to explain which lorry:

(a) reaches the greatest velocity ....

.... *(2 marks)*

(b) achieves the greatest acceleration ....

.... *(2 marks)*

B-A* **3** The velocity–time graph shows part of the journey of a train.

EQA SKILL Interpret Page 121

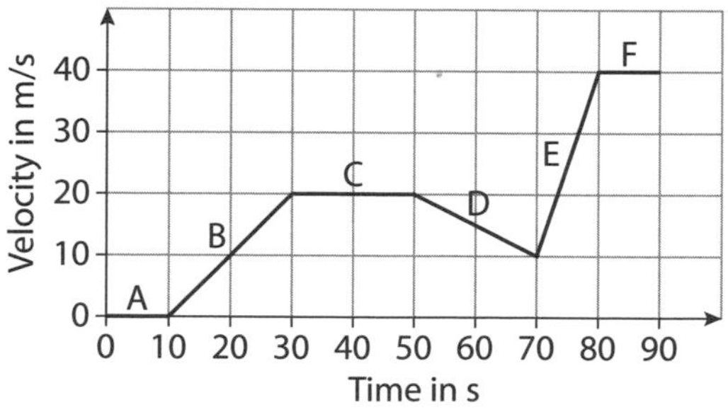

(a) Calculate the acceleration of the train in part B of its journey.

....

....

....

Acceleration = ... m/s² *(2 marks)*

(b) Calculate the acceleration in part D of the journey.

....

....

Acceleration = ... m/s² *(2 marks)*

(c) Calculate the distance travelled between 10 and 50 seconds.

....

....

Distance = ... m *(2 marks)*

Forces and braking

1 As part of a driving test in the UK, drivers are expected to perform an 'emergency stop'. This means they need to stop in as short a distance as possible, without skidding.

The table below shows some data for three of these emergency stops that have been performed in recent driving tests.

Speed in miles per hour	Braking force in N	Braking distance in m
20	10 000	6
40	10 000	24
60	10 000	54

(a) Describe how braking distance varies with speed.

....

.... *(2 marks)*

(b) What would need to happen to the braking force of a car travelling at 40 mph to enable the car to be brought to rest in 20 m?

.... *(1 mark)*

Guided

(c) Explain why the brakes become hot as the car slows down.

The friction between the brakes and the wheels reduces the ....

energy of the vehicle. This energy is transferred to the brakes, which ....

.... *(2 marks)*

D-C

2 **(a)** State the meaning of **thinking distance**.

.... *(1 mark)*

(b) State **two** factors that can increase the thinking distance.

1 ....

2 ....

(2 marks)

(c) State **two** factors that can increase the braking distance.

1 ....

2 ....

(2 marks)

B-A*

3 **(a)** Explain why the factors that affect the thinking distance of a moving vehicle at a particular speed are all connected with the driver, not the vehicle or the state of the road.

....

.... *(2 marks)*

(b) The factors that affect the braking distance at a particular speed are usually all connected with the vehicle or the road. Suggest how the condition of the driver could also affect the braking distance.

....

.... *(2 marks)*

Had a go ☐ Nearly there ☐ Nailed it! ☐

Falling objects

In an exam, all the equations you need will be given to you on a separate sheet. At Higher Tier you are expected to be able to rearrange equations when necessary.

For each of these questions, write down the equation you use, and then show clearly how you work out your answer.

D-C 1 On Earth, the value for the gravitational field strength is 10 N/kg.

(a) Calculate the weight of a man of mass 76 kg.

....

Weight = ... N *(2 marks)*

Guided **(b)** An apple has a mass of 80 g. Calculate its weight.

mass in kg = $\frac{80\ g}{1000}$ = ...

....

....

Weight = ... N *(3 marks)*

B-A* 2 **(a)** Submarines can travel at different speeds underwater. Describe the relationship between the speed of the submarine and the size of the drag force from the water.

.... *(1 mark)*

(b) A submarine weighs 200 000 000 N. Calculate its mass.

....

....

Mass = ... kg *(2 marks)*

B-A* 3 A skydiver jumps out of an aeroplane. She falls until she reaches terminal velocity, and then she opens her parachute. She continues to fall until she reaches the ground.
Explain how the following forces change during her jump.

(a) Weight.

.... *(1 mark)*

(b) Air resistance.

....

....

....

.... *(4 marks)*

Forces and terminal velocity

D-C

1 A skydiver jumps out of a plane from 3000 m and falls towards the ground. The velocity–time graph shows how his velocity changes.

EQA SKILL Interpret Page 121

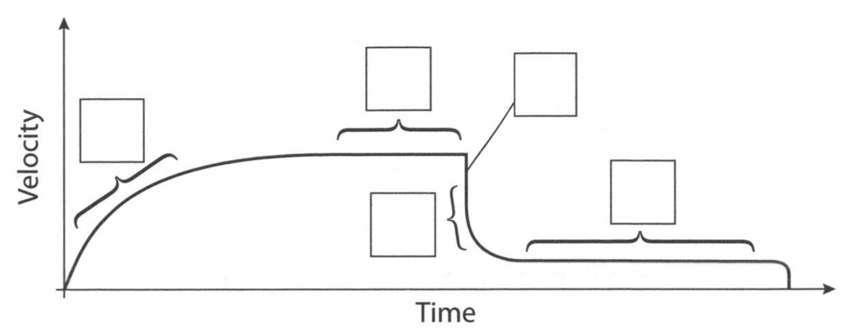

Add the following labels to the graph.

(a) A – resultant force is zero *(2 marks)*

There are two marks, so you need to label two points on the graph.

(b) B – greatest acceleration *(1 mark)*

(c) C – air resistance increasing because speed is increasing *(1 mark)*

(d) D – air resistance decreasing because speed is decreasing *(1 mark)*

B-A*

2 Dragsters are racing cars that race over a very short course. The cars have very large accelerations. The race involves finding the car that can cover the length of the racetrack in the shortest time.

EQA SKILL Interpret Page 121

Exam questions are often set in different contexts. You may not have studied drag racing, but you should be able to use your knowledge of air resistance and speed to answer the question.

One dragster reaches a velocity of 400 kilometres per hour in only 4 seconds. The car crosses the finishing line 7 seconds after starting the race, and then releases a parachute to help it to slow down.

(a) **(i)** Sketch a velocity–time graph to show how the dragster's velocity changes over the course of the race. *(4 marks)*

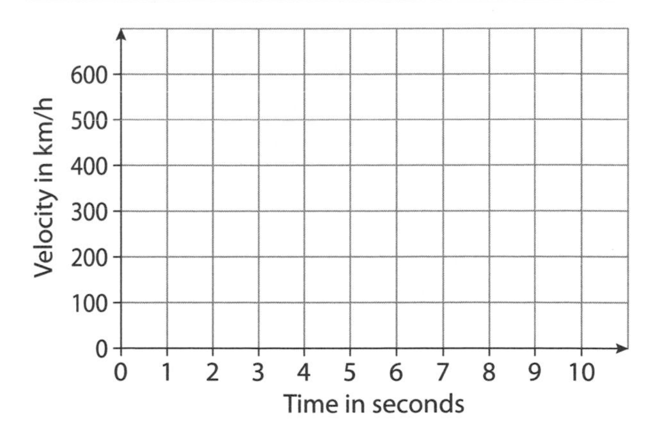

Guided

(ii) Explain the shape of the section of your graph from 0 to 3 seconds.

When the car is just starting the air resistance is ....

and so the resultant force ....

....

....

.... *(4 marks)*

(b) On your graph, write the letter of the statement below in the place that best describes it.

A – greatest forwards acceleration

B – terminal velocity reached

C – parachute opens *(3 marks)*

Had a go ☐ Nearly there ☐ Nailed it! ☐

Elasticity

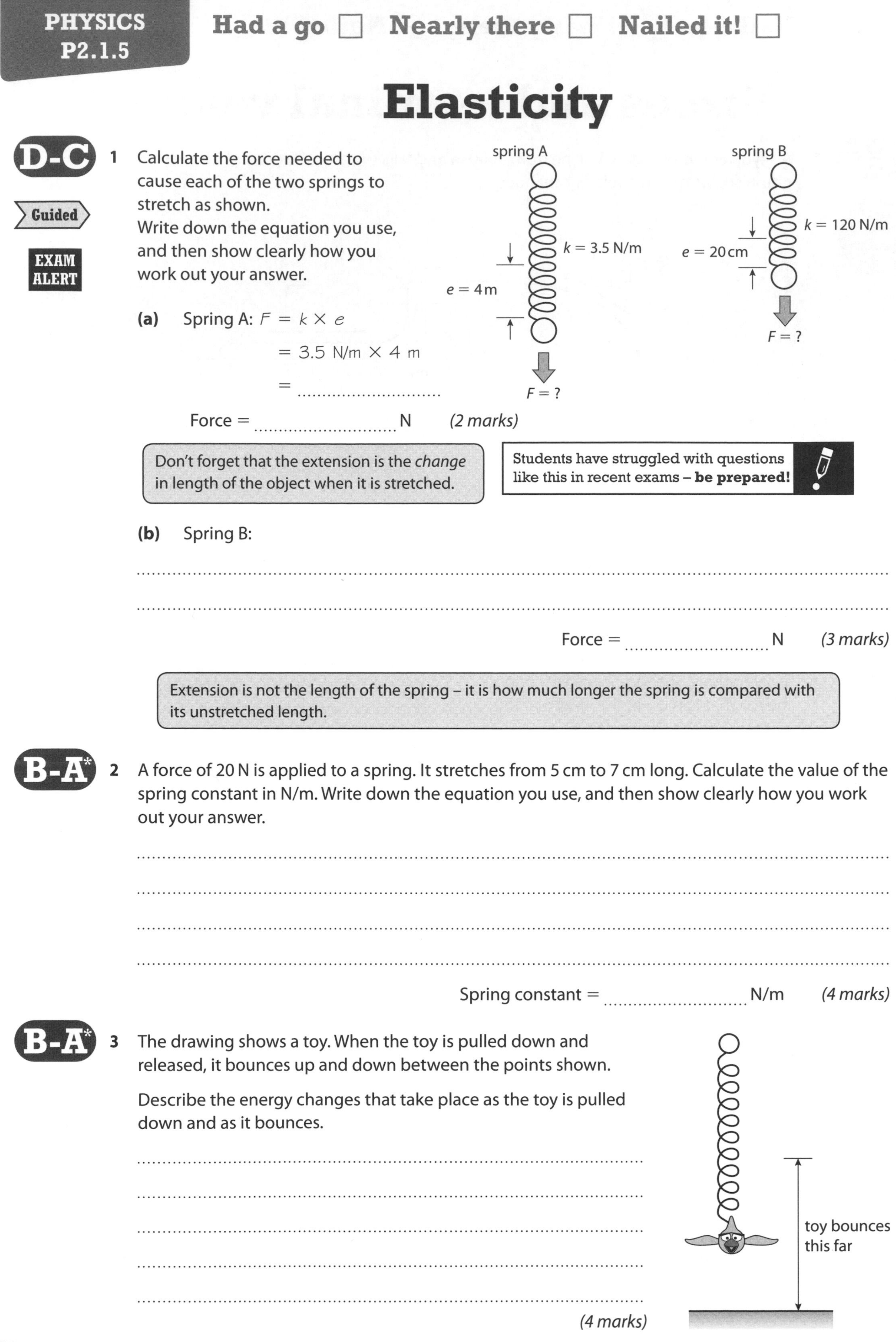

D-C

Guided

EXAM ALERT

1 Calculate the force needed to cause each of the two springs to stretch as shown. Write down the equation you use, and then show clearly how you work out your answer.

(a) Spring A: $F = k \times e$

$= 3.5$ N/m $\times$ 4 m

= ....

Force = ... N *(2 marks)*

Don't forget that the extension is the *change* in length of the object when it is stretched.

Students have struggled with questions like this in recent exams – **be prepared!**

(b) Spring B:

....

....

Force = ... N *(3 marks)*

Extension is not the length of the spring – it is how much longer the spring is compared with its unstretched length.

B-A*

2 A force of 20 N is applied to a spring. It stretches from 5 cm to 7 cm long. Calculate the value of the spring constant in N/m. Write down the equation you use, and then show clearly how you work out your answer.

....

....

....

....

Spring constant = ... N/m *(4 marks)*

B-A*

3 The drawing shows a toy. When the toy is pulled down and released, it bounces up and down between the points shown.

Describe the energy changes that take place as the toy is pulled down and as it bounces.

....

....

....

....

....

(4 marks)

Forces and energy

For all calculation questions, write down the equation you use, and then show clearly how you work out your answer.

D-C **1** The diagram shows a wheelbarrow being pushed up a ramp. Explain which force and distance should be used to calculate the work done in pushing the barrow.

....

....

....

.... *(3 marks)*

D-C **2** A woman pushes a heavy box 5 metres along the floor. She needs a force of 50 N to move the box. Calculate the work done in pushing the box. Give the unit for work done.

Guided

$W = F \times d$

$= 50\ N \times 5\ m$

= ....

Work done = ... Unit .... *(3 marks)*

D-C **3** A man uses 20 J of energy to lift a pot of paint onto a shelf. It takes him 2 seconds to lift the paint. Calculate the power. Give the unit in your answer.

....

....

Power = ... Unit .... *(3 marks)*

B-A* **4** A woman uses 150 J of energy to lift a crate onto a shelf 50 cm above the floor. Calculate the weight of the box.

....

....

....

Weight = ... N *(3 marks)*

B-A* **5** The brakes of a car exert a force of 8000 N to bring it to a stop in 4 seconds. The car covers a distance of 25 metres while braking. Calculate the power of the brakes.

You may get some marks for the working even if you get the final answer wrong. This is particularly important in questions like this, where you need to carry out two calculations.

....

....

....

....

Power = ... W *(4 marks)*

Had a go ☐ Nearly there ☐ Nailed it! ☐

KE and GPE

For all calculation questions, write down the equation you use, and then show clearly how you work out your answer. The value of g is 10 N/kg.

D-C

1 The mass of a vase is 0.8 kg. Calculate the gravitational potential energy it gains when it is placed on a shelf that is 1.5 m above the ground.

....

....

GPE = ... J *(2 marks)*

D-C

2 A ball has a mass of 180 g. It is thrown up to 4 m in the air.
Calculate the gravitational potential energy the ball has at the top of the throw.

....

....

....

GPE = ... J *(3 marks)*

D-C

Guided

EXAM ALERT

3 A bowling ball has a mass of 4 kg. It is rolling at a speed of 1.5 m/s.
Calculate its kinetic energy.

$E_k = \frac{1}{2} \times m \times v \times v$

$= 0.5 \times 4\,kg \times 1.5\,m/s \times 1.5\,m/s$

= ....

Don't forget that v^2 means speed squared, or $v \times v$

Students have struggled with questions like this in recent exams – **be prepared!**

KE = ... J *(2 marks)*

D-C

4 Explain how crumple zones in cars help prevent injuries to the occupants of a car in an accident.

....

....

.... *(3 marks)*

B-A*

5 A book with a mass of 1 kg falls off a shelf that is 2 m above the floor.
Calculate its velocity when it hits the floor.

First calculate the gravitational potential energy the book had when it was on the shelf.

....

....

....

....

Velocity = ... m/s *(5 marks)*

Momentum

For all calculation questions, write down the equation you use, and then show clearly how you work out your answer.

D-C **1** A hockey ball with a mass of 160 g is moving at 2 m/s. Calculate its momentum.

Guided

160 g = 0.16 kg

$p = m \times v = 0.16\ \text{kg} \times 2\ \text{m/s} =$ ....

Momentum = ... kg m/s *(3 marks)*

D-C **2** Explain how the momentum of a moving object would change if the direction of motion was reversed.

....

.... *(2 marks)*

B-A* **3** In an experiment, a 1.5 kg trolley moving at 0.2 m/s collides with a 1 kg trolley moving at 0.1 m/s in the same direction. The two trolleys stick together.

(a) Calculate the total momentum of the trolleys before the collision.

....

....

Momentum = ... kg m/s *(2 marks)*

(b) Calculate the velocity of the combined trolleys after the collision.

....

....

....

Velocity = ... m/s *(3 marks)*

B-A* **4** A bomb splits into two pieces when it explodes.

(a) Explain how it is possible for its momentum before it explodes to be equal to its momentum after it has exploded.

....

....

.... *(3 marks)*

(b) One piece of the bomb has a mass of 4 kg. It flies off to the north at a speed of 240 m/s. The other piece of the bomb has a mass of 6 kg. Calculate the velocity of the 6 kg piece of the bomb.

> Start by calculating the momentum of the 4 kg piece. You can then work out the momentum and hence the velocity of the remaining piece.

....

....

....

....

Velocity = ... m/s *(5 marks)*

Physics six mark question 1

Modern cars can be driven at high speeds, which means that a crash could kill or injure the driver and passengers. However, modern cars are designed with many safety features that help the car to stop suddenly and protect the occupants if there is a crash. These safety features include crumple zones, air bags, seat belts and side impact bars.

Explain how some of these safety features work.

You will be more successful in six mark questions if you plan your answer before you start writing.

Think about the safety features fitted to cars and relate these to the energy changes involved when a car crashes.

You could include information on the following:

- How is the kinetic energy of a moving car normally reduced?
- How is the kinetic energy of a moving car reduced in a crash?
- What do crumple zones and air bags do?
- Why should people in a car wear seat belts?
- How do side impact bars work?

(6 marks)

Static electricity

D-C **1** The diagram shows two rods made of an insulating material. The polythene rod has a negative charge.

X

polythene

Guided **(a)** The polythene rod was given a charge by rubbing it with a cloth. Explain what happened to give the polythene rod its charge.

Electrons were .... from the .... to the

....

(2 marks)

(b) Rod X is also given a negative charge. Explain what will happen when the two rods are hung next to each other, as shown in the diagram.

....

.... *(2 marks)*

(c) Rod X is replaced with a rod that has a positive charge. A student says that the rod has a positive charge because protons were transferred to it when it was rubbed. Explain why the student is wrong.

....

.... *(2 marks)*

D-C **2** Sometimes combing your hair can make bits of it stick out. Explain why this happens.

....

....

....

.... *(4 marks)*

B-A* **3** Two students are discussing static electricity.

A: Rubbing a metal does not transfer electrons.

B: Rubbing any material can transfer electrons.

Explain which student is correct.

....

....

....

.... *(3 marks)*

Had a go ☐ Nearly there ☐ Nailed it! ☐

Current and potential difference

For each of these questions, write down the equation you use, and then show clearly how you work out your answer.

D-C **1** **(a)** 600 C of electric charge flow through a component in 40 seconds. Calculate the current in the circuit.

$I = \frac{Q}{t} = \frac{600\ C}{40\ s} =$ .... Current = ... A *(2 marks)*

(b) The charge does 240 000 J of work as it flows through the component. Calculate the potential difference across the component.

....

.... *(2 marks)*

B-A* **2** The graph shows how the current in a hair dryer changes over a period of 6 minutes. Use the graph to calculate the following.

Current in A
10
8
6
4
2
0
0 1 2 3 4 5 6
Time in minutes

(a) Calculate the total charge that flowed through the hair dryer during the 6 minutes it was in use.

....

....

....

....

....

Charge = ... C *(4 marks)*

(b) The hair dryer is plugged into a 230 V supply. Calculate the energy transferred by the hair dryer during the 6 minutes.

....

....

Energy = ... J *(2 marks)*

(c) In some countries the mains voltage is 110 V.

(i) Calculate the charge that flows through the hair dryer when the 110 V supply transfers 200 000 J of energy.

....

....

Charge = ... C *(2 marks)*

(ii) This energy is transferred in 6 minutes. Calculate the current in the hair dryer.

....

....

Current = ... A *(2 marks)*

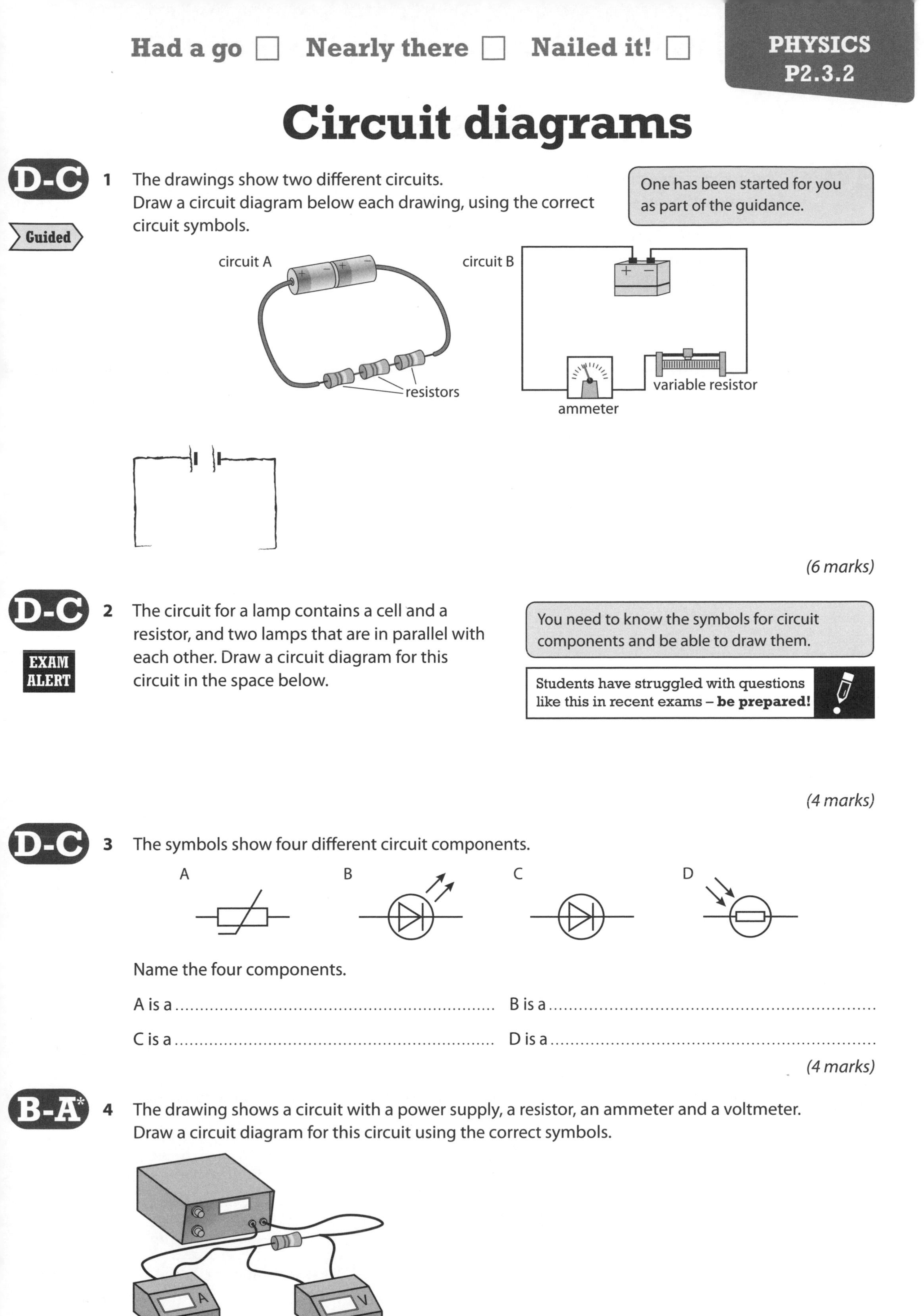

Circuit diagrams

D-C **1** The drawings show two different circuits.
Draw a circuit diagram below each drawing, using the correct circuit symbols.

Guided

One has been started for you as part of the guidance.

(6 marks)

D-C **2** The circuit for a lamp contains a cell and a resistor, and two lamps that are in parallel with each other. Draw a circuit diagram for this circuit in the space below.

EXAM ALERT

You need to know the symbols for circuit components and be able to draw them.

Students have struggled with questions like this in recent exams – **be prepared!**

(4 marks)

D-C **3** The symbols show four different circuit components.

Name the four components.

A is a .... B is a ....

C is a .... D is a ....

(4 marks)

B-A* **4** The drawing shows a circuit with a power supply, a resistor, an ammeter and a voltmeter.
Draw a circuit diagram for this circuit using the correct symbols.

(2 marks)

Resistors

D-C **1** The graph shows the relationship between current and potential difference for two resistors.

(a) Explain how you can tell from the graph that resistor A has a greater resistance than resistor B, without carrying out any calculations.

....

....

(2 marks)

(b) Use the graph to find the current flowing in resistor A when the potential difference is 2.5 V.

.... *(1 mark)*

D-C **Guided** **2** A current of 1.5 A flows through a resistor of resistance 800 Ω. Calculate the potential difference across the resistor. Write down the equation you use, and then show clearly how you work out your answer.

$V = I \times R$

$= 1.5\text{ A} \times 800\,\Omega$

= ....

Potential difference = ... V *(2 marks)*

B-A* **3** For each of these questions, write down the equation you use, and then show clearly how you work out your answer.

(a) A potential difference of 12 000 V is applied across a component. The resistance of the component is 1000 Ω. Calculate the current.

....

....

Current = ... A *(2 marks)*

(b) A potential difference of 12 V causes a current of 0.01 A to flow through a wire. Calculate the resistance of the wire.

....

....

Resistance = ... Ω *(2 marks)*

B-A* **4** Use the graph in question **1** to determine the resistance of resistor A and the resistance of resistor B.

Resistor A

....

....

....

Resistance = ... Ω

Resistor B

....

....

....

Resistance = ... Ω

(4 marks)

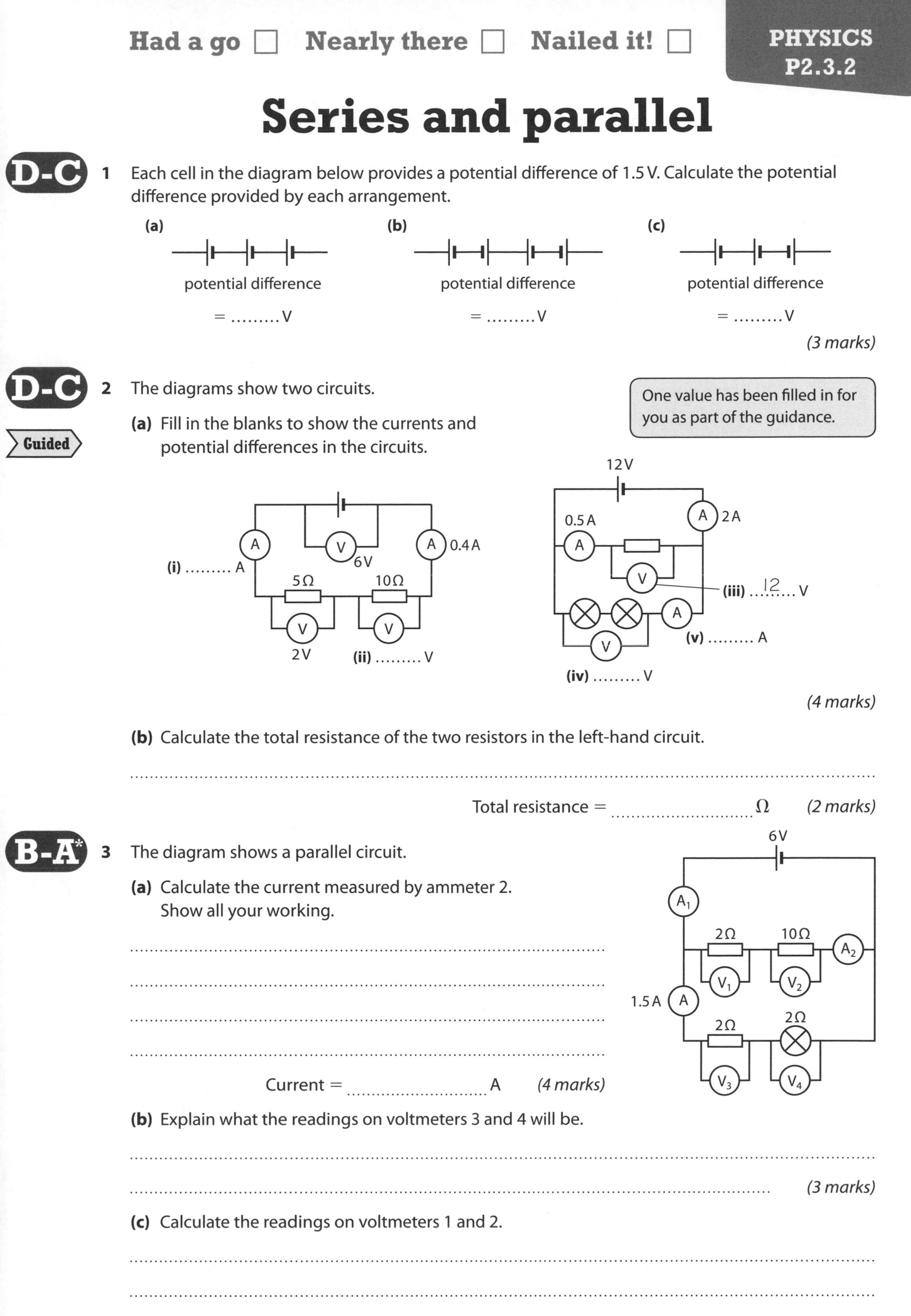

Series and parallel

D-C **1** Each cell in the diagram below provides a potential difference of 1.5 V. Calculate the potential difference provided by each arrangement.

(a) potential difference = ... V

(b) potential difference = ... V

(c) potential difference = ... V

(3 marks)

D-C **Guided** **2** The diagrams show two circuits.

One value has been filled in for you as part of the guidance.

(a) Fill in the blanks to show the currents and potential differences in the circuits.

(4 marks)

(b) Calculate the total resistance of the two resistors in the left-hand circuit.

....

Total resistance = ... Ω *(2 marks)*

B-A* **3** The diagram shows a parallel circuit.

(a) Calculate the current measured by ammeter 2.
Show all your working.

....

....

....

....

Current = ... A *(4 marks)*

(b) Explain what the readings on voltmeters 3 and 4 will be.

....

.... *(3 marks)*

(c) Calculate the readings on voltmeters 1 and 2.

....

....

.... *(3 marks)*

Had a go ☐ Nearly there ☐ Nailed it! ☐

Variable resistance

D-C

1 (a) Name a component that can be used to control a circuit that closes curtains in a house when it goes dark outside. .... *(1 mark)*

Guided

(b) Explain why you chose this component.

The resistance .... when brighter light shines on it.

(2 marks)

D-C

2 An electrical component will be damaged if the battery in its circuit is inserted the wrong way round.

(a) Name **one** component that can be used to protect other components in the circuit by stopping the current flowing if the battery is not inserted correctly.

.... *(1 mark)*

(b) Draw a line on the graph to show how the current varies with potential difference in the component you have named.

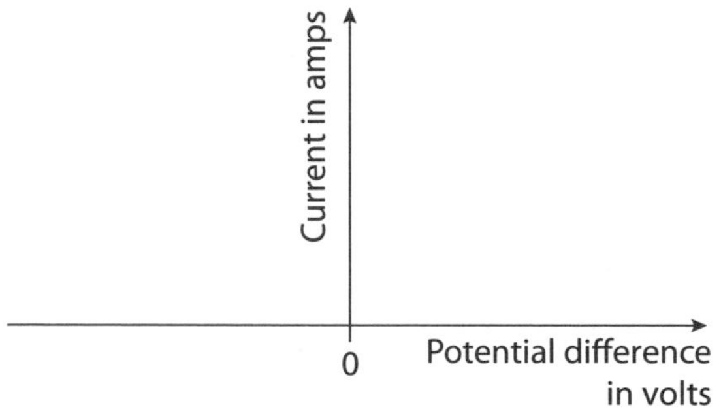

Current in amps

0

Potential difference in volts

(3 marks)

D-C

3 (a) Describe how the resistance of a thermistor can be increased.

.... *(1 mark)*

(b) Suggest **one** use for a thermistor.

....

.... *(1 mark)*

B-A*

4 The graph shows how the current through a filament bulb changes as the potential difference across the bulb changes.

Current in amps

Potential difference in volts

Explain the shape of the graph. Your answer should include why the resistance changes as well as how it changes.

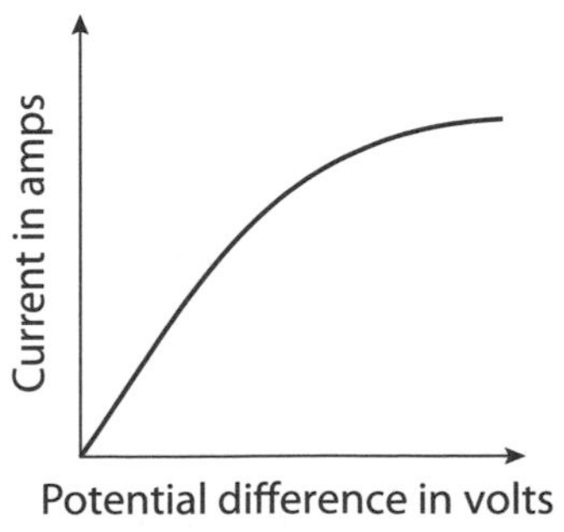

....

....

....

....

....

(4 marks)

Using electrical circuits

D-C

1 The table shows information about three different types of lighting – a filament bulb, a fluorescent bulb and a light-emitting diode (LED).

Comparisons are for bulbs with the same light output	LED bulb	Filament bulb	Compact fluorescent bulb
Electrical power used	8 watts	60 watts	12 watts
Average life	50 000 hours	1200 hours	8000 hours
Cost	£14.40	£1.50	£2.90
Disposal	Can put in normal waste	Can put in normal waste	Contains mercury, must be disposed of properly

(a) The government is reducing the number of filament bulbs that can be sold. Suggest what the environmental impact of filament bulbs might be.

Filament bulbs use a lot more .... than other types of bulb for the same .... Most electricity is generated using ....

....

(3 marks)

(b) Evaluate LED bulbs and compact fluorescent bulbs for use in homes.

> Evaluate means you have to give some advantages and disadvantages of each, and give your opinion about which is the best.

....

....

....

.... *(4 marks)*

B-A*

2 Look at the circuit below. The LED emits light when the potential difference across it is 1 V. When the thermistor is cold its resistance is high. As the potential difference from the cell is shared between the components, the thermistor has a higher potential difference across it than the LED.

6V

(a) Explain why the LED does not light up when the circuit is cold.

....

.... *(2 marks)*

(b) Explain what will happen if the temperature of the thermistor is increased.

....

.... *(3 marks)*

Had a go ☐ Nearly there ☐ Nailed it! ☐

Different currents

1 **(a)** Describe the difference between direct and alternating current.

In a direct current, the current always flows ....

.... *(2 marks)*

(b) State the potential difference and frequency of the mains supply in the UK.

Potential difference = ... V

Frequency = ... Hz *(2 marks)*

(c) Name **one** component that supplies a direct current. .... *(1 mark)*

2 The oscilloscope is showing how the potential difference of an alternating supply changes. The vertical scale is 2 V per square.

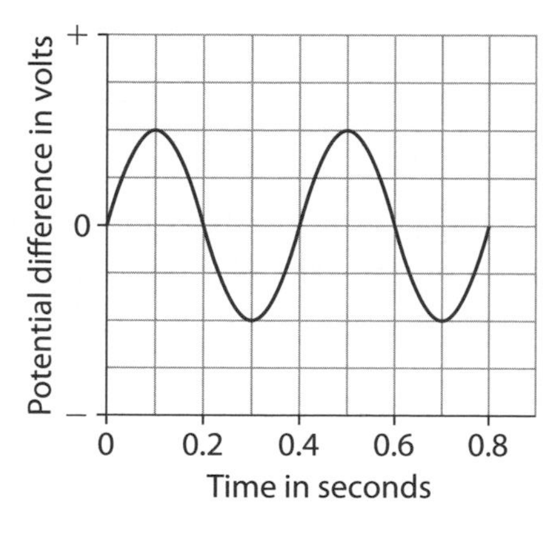

(a) How many complete cycles does the oscilloscope show? .... *(1 mark)*

(b) What is the peak potential difference? .... *(1 mark)*

(c) Draw a line on the graph to show a 6 V d.c. supply. *(2 marks)*

3 Look at the oscilloscope trace in question **2**.

(a) What is the period of this a.c. supply? .... *(1 mark)*

(b) Calculate the frequency of the supply. Write down the equation you use, and then show clearly how you work out your answer.

You need to remember that frequency = 1/period. This equation will not be given to you in the exam.

....

....

Frequency = ... Hz *(2 marks)*

(c) Calculate the period of a 30 Hz supply. Write down the equation you use, and then show clearly how you work out your answer.

....

....

Period = ... s *(2 marks)*

(d) A 2 Hz supply has a period of 0.5 s and a peak potential difference of 3 V. Sketch a curve on the graph in question **2** to show this supply. *(3 marks)*

Three-pin plugs

D-C

1 Many electrical appliances have a three-core cable connected to a three-pin plug.

(a) Describe the structure of a three-core cable.

Guided

A three-core cable consists of three metal wires. Each wire is ....

....

.... (3 marks)

(b) Explain why:

(i) copper is used for the wires.

.... (1 mark)

(ii) a flexible plastic is used for the outer coverings.

....

.... (2 marks)

D-C

2 The three wires inside a three-core cable have differently coloured coatings.

(a) Complete the table to show the colours used.

Wire	Colour
live	
neutral	
earth	

(3 marks)

(b) A two-core cable only has two wires.
Name the wire that is not included in a two-core cable. .... (1 mark)

(c) Explain why the wires have coloured coatings.

....

.... (2 marks)

B-A*

3 The table below shows the relative properties of four different metals. Use data from the table to explain why the wires in an electrical cable are made from copper but the pins in a three-pin plug are made from brass.

Material	Conductivity	Cost	Strength
copper	***	£££	***
brass	**	£££	****

Remember to mention why each material might not be a good choice for each use as well as why it is useful.

(More stars means better conductivity or higher strength.)

....

....

....

.... (4 marks)

Had a go ☐ Nearly there ☐ Nailed it! ☐

Electrical safety

D-C

1 A three-pin plug includes a fuse.

Guided

(a) Describe how a fuse works.

If the current gets too high the fuse ....

.... *(3 marks)*

(b) Explain why a fuse or a circuit breaker should always be included in a mains circuit.

.... *(1 mark)*

(c) (i) Describe how a **residual current circuit breaker** (RCCB) works.

....

.... *(2 marks)*

(ii) State **one** advantage of an RCCB compared with a fuse.

.... *(1 mark)*

D-C

2 Some appliances are described as **double-insulated**.

(a) What does this tell you about the appliance?

.... *(1 mark)*

(b) How is the cable used for a double-insulated appliance different from the cable for an appliance that is not double-insulated?

.... *(1 mark)*

B-A*

AQA SKILL Explain Page 121

3 The electrical circuits in a house all start from a consumer unit. There are usually separate circuits for lights and for power sockets. If the home has an electric cooker or electric shower, each of these items usually has its own circuit. Each circuit has its own separate circuit breaker.

(a) Suggest why there are usually lots of different circuits in a home.

.... *(1 mark)*

(b) A consumer unit usually contains circuit breakers with different current ratings. Explain why this is so.

....

....

.... *(3 marks)*

(c) The wires used in the different circuits are usually of different thicknesses. Explain why this is done.

Think about why some circuits may need thicker wires than others, and also consider why not all the circuits are made of the thickest wires.

....

....

....

....

.... *(4 marks)*

Current and power

For each of these questions, write down the equation you use, and then show clearly how you work out your answer.

D-C **1** A light bulb uses the mains electricity supply (230 V). It transfers 72 000 J of energy in 1 hour. Calculate the power of the bulb.

Guided

1 hour = 60 × 60 = ... seconds

$P = \frac{E}{t}$

= ....

Power = ... W *(3 marks)*

D-C **2** A battery-powered torch uses a 3 V battery. When it is switched on the current in the bulb is 1.5 A. Calculate the power of the bulb.

....

....

Power = ... W *(2 marks)*

D-C **3** State why 'low-energy' lamps such as compact fluorescent lamps are more efficient than filament bulbs.

....

....

(1 mark)

B-A* **4** A defibrillator is used to restart the heart of somebody who has suffered a heart attack. Typical values for a defibrillator are shown in the box.

Potential difference output = 1500 V
Time for which defibrillator is used on body = 0.01 s
Energy delivered by defibrillator = 120 J

(a) Calculate the power of the defibrillator.

....

....

Power = ... W *(2 marks)*

(b) Calculate the current through the patient's chest.

....

....

....

Current = ... A *(2 marks)*

(c) Calculate the charge that flows during the time for which the defibrillator is used on the body.

....

....

Charge = ... C *(2 marks)*

Physics six mark question 2

A student uses a cloth to rub two polythene rods, a Perspex rod and a metal rod. Polythene gains a negative charge and Perspex becomes positively charged. The rods are suspended from threads.

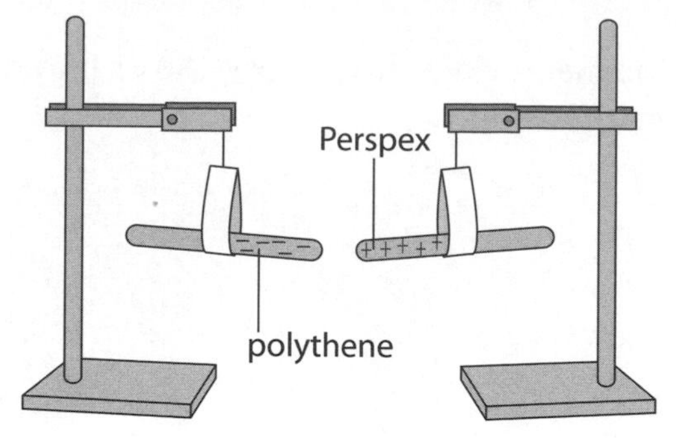

Explain how the rods become charged, and what will happen when different pairs of rods are suspended next to each other.

You will be more successful in six mark questions if you plan your answer before you start writing.

You could include information on the following:

- which particles are transferred when an object gets a positive or negative charge of static electricity
- how two charged objects will attract or repel one another
- why a metal rod will not get a static charge.

....

....

....

....

....

....

....

....

....

....

....

....

....

....

....

....

....

....

....

.... *(6 marks)*

Atomic structure

1 The table shows the atomic structure of some different particles.

Particle	Number of protons	Number of neutrons	Number of electrons
A	8	8	8
B	9	10	9
C	11	10	10
D	8	10	9
E	10	10	10

(a) State the atomic number and mass number of particle C.

Atomic number = ... Mass number = ... *(2 marks)*

(b) Explain which **two** particles are isotopes of the same element.

Particles A and D are isotopes of the same element because they have the same

.... but different

.... *(3 marks)*

(c) (i) Explain which **two** particles are ions.

....

.... *(2 marks)*

(ii) Explain which of these particles is a positive ion.

....

.... *(2 marks)*

2 The table shows some properties of the particles that make up atoms.
Complete the missing information in the table.

Particle	Charge	Relative mass	Position in atom
proton		1	
neutron	0		
electron		Very small	In orbit around nucleus

(5 marks)

3 The mass number of an individual atom is a whole number. However, a copy of the Periodic Table shows chlorine with an atomic number of 17 and a mass number of 35.5.

Explain what the information on this periodic table tells us about the atomic structure of chlorine atoms.

There are three marks for this question, so you need to make three separate points in your answer.

....

....

.... *(3 marks)*

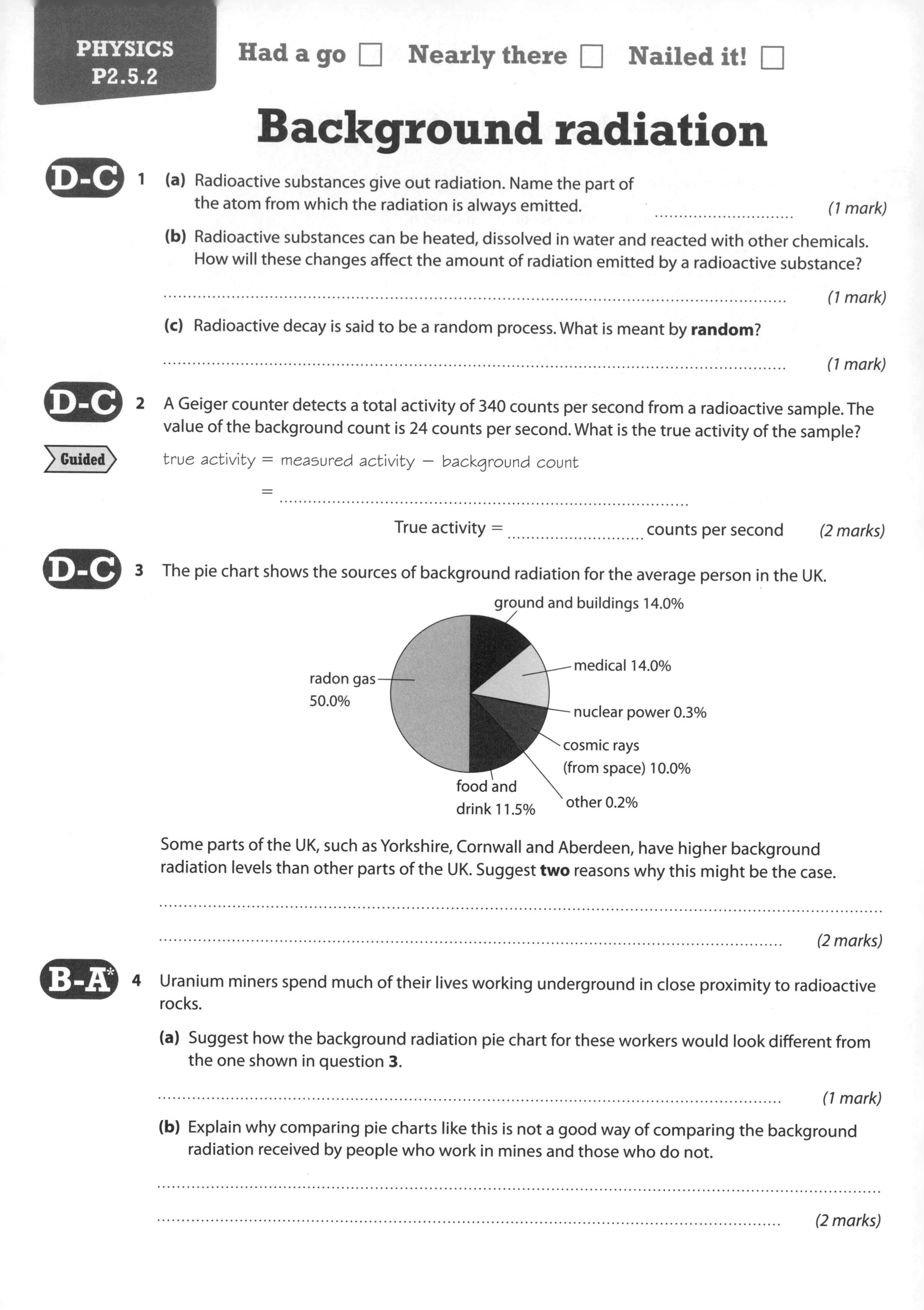

PHYSICS P2.5.2

Had a go ☐ Nearly there ☐ Nailed it! ☐

Background radiation

D-C 1 **(a)** Radioactive substances give out radiation. Name the part of the atom from which the radiation is always emitted. .... *(1 mark)*

(b) Radioactive substances can be heated, dissolved in water and reacted with other chemicals. How will these changes affect the amount of radiation emitted by a radioactive substance?

.... *(1 mark)*

(c) Radioactive decay is said to be a random process. What is meant by **random**?

.... *(1 mark)*

D-C 2 A Geiger counter detects a total activity of 340 counts per second from a radioactive sample. The value of the background count is 24 counts per second. What is the true activity of the sample?

Guided true activity = measured activity − background count

= ....

True activity = ... counts per second *(2 marks)*

D-C 3 The pie chart shows the sources of background radiation for the average person in the UK.

Some parts of the UK, such as Yorkshire, Cornwall and Aberdeen, have higher background radiation levels than other parts of the UK. Suggest **two** reasons why this might be the case.

....

.... *(2 marks)*

B-A* 4 Uranium miners spend much of their lives working underground in close proximity to radioactive rocks.

(a) Suggest how the background radiation pie chart for these workers would look different from the one shown in question **3**.

.... *(1 mark)*

(b) Explain why comparing pie charts like this is not a good way of comparing the background radiation received by people who work in mines and those who do not.

....

.... *(2 marks)*

Nuclear reactions

D-C **1** The table shows three types of radiation that can be given off by radioactive nuclei.

(a) Complete the missing information in the table.

Radiation	Description	Charge	Mass number	Atomic number
alpha	two protons and two neutrons			2
beta		−1		−1
gamma	electromagnetic radiation			

(4 marks)

Guided **(b)** Explain why the following symbol is used to represent an alpha particle: $^{4}_{2}He$

A helium nucleus has two .... and two ....,

which is the same ....

(2 marks)

(c) An isotope of radium has 88 protons and 138 neutrons.
Add the correct numbers to this symbol for radium.

$^{\square}_{\square}Ra$ *(2 marks)*

B-A* **2** Complete the nuclear equations for the following nuclear reactions.

(a) Plutonium-239 decays by emitting an alpha particle.

$^{239}_{94}Pu \rightarrow {}^{\square}_{\square}U + {}^{4}_{2}He$ *(2 marks)*

(b) Technetium-99 decays by emitting a beta particle.

$^{\square}_{43}Tc \rightarrow {}^{\square}_{44}Ru + {}^{0}_{\square}e$ *(3 marks)*

B-A* **3** Many radioactive isotopes decay to form another radioactive isotope. This isotope decays in its turn.

The diagram below shows part of the decay chain for neptunium-237. Complete the missing information to show the type of particle emitted at each stage.

> You do not need to know about decay chains for your exam. You can answer this question from your knowledge of nuclear equations.

$^{237}_{93}Np \rightarrow {}^{233}_{91}Pa \rightarrow {}^{233}_{92}U \rightarrow {}^{229}_{90}Th \rightarrow {}^{225}_{88}Ra \rightarrow {}^{225}_{89}Ac$

↘ α ↘ β ↘ .... ↘ .... ↘ ....

(3 marks)

PHYSICS
P2.5.2

Had a go ☐ Nearly there ☐ Nailed it! ☐

Alpha, beta and gamma radiation

D-C **1** The diagram below shows α, β and γ radiation from a radioactive source being deflected by a magnetic field. Label each radiation correctly.

EXAM ALERT

(b) ....

(a) ....

(c) ....

You need to learn the properties of the three types of radiation.

Students have struggled with questions like this in recent exams – **be prepared!**

(3 marks)

D-C **2** Alpha radiation is relatively safe if the source is outside the body, but is very harmful if the source is swallowed or breathed in. Explain why this is so.

Guided

Alpha particles are not very penetrating so they cannot

AQA SKILL Explain Page 121

....

....

.... *(4 marks)*

B-A* **3** A radioactive source emits alpha and beta particles. The stream of particles is passed through an electric field.

(a) State why alpha and beta particles are deflected in opposite directions.

.... *(1 mark)*

(b) Explain why the alpha and beta particles are deflected by different amounts.

There are 3 marks for this part of the question, so you need to make three points in your answer.

....

....

.... *(3 marks)*

(c) Explain how the electric field would affect gamma radiation.

....

.... *(2 marks)*

Half-life

D-C

1 The graph shows how the count rate from a radioactive sample changes with time. Use the graph below to determine half-life of the sample.

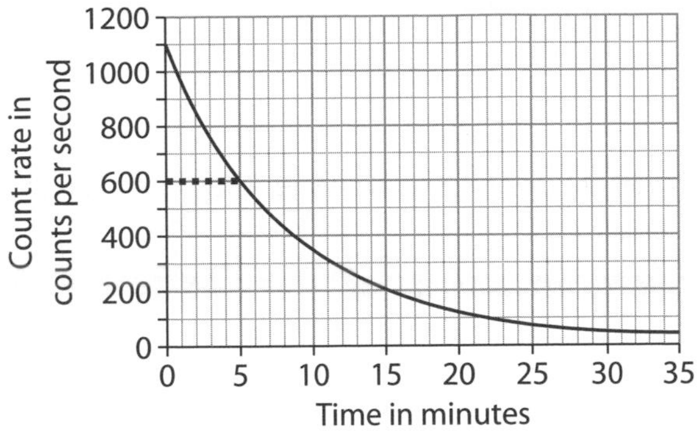

Guided

Count rate in counts per second: 0, 200, 400, 600, 800, 1000, 1200

Time in minutes: 0, 5, 10, 15, 20, 25, 30, 35

It is easier to answer this kind of question if you draw lines on the graph. The horizontal dashed line on the graph has been drawn for you as part of the guidance.

The count rate is 600 per second at 5 minutes.

It is half of this (300) at .... minutes.

Half-life = ... − 5 minutes = ...

You can start at any point on the graph.

Half-life = ... minutes *(2 marks)*

D-C

2 The activity of a sample is 1000 counts per minute, and it has a half-life of 10 minutes. What is its activity 30 minutes later?

....

....

....

Activity = ... counts per minute *(3 marks)*

B-A*

3 The following information is about an element called technetium, which is used as a tracer in hospitals.

AQA SKILL Describe Page 121

> Technetium-99 is a silver-grey, radioactive metal. It occurs naturally only in very small amounts. Technetium-99 has a radioactive half-life of 212 000 years and decays to form ruthenium-99, which is stable, by emitting beta and gamma radiation. Technetium-99m (called metastable Tc-99) has a half-life of only about 6 hours and decays to Tc-99 primarily by gamma emission.

(a) Describe how a radioactive tracer is used in medicine.

....

....

.... *(3 marks)*

(b) Explain which form of Tc-99 is used as a tracer.

....

....

....

.... *(4 marks)*

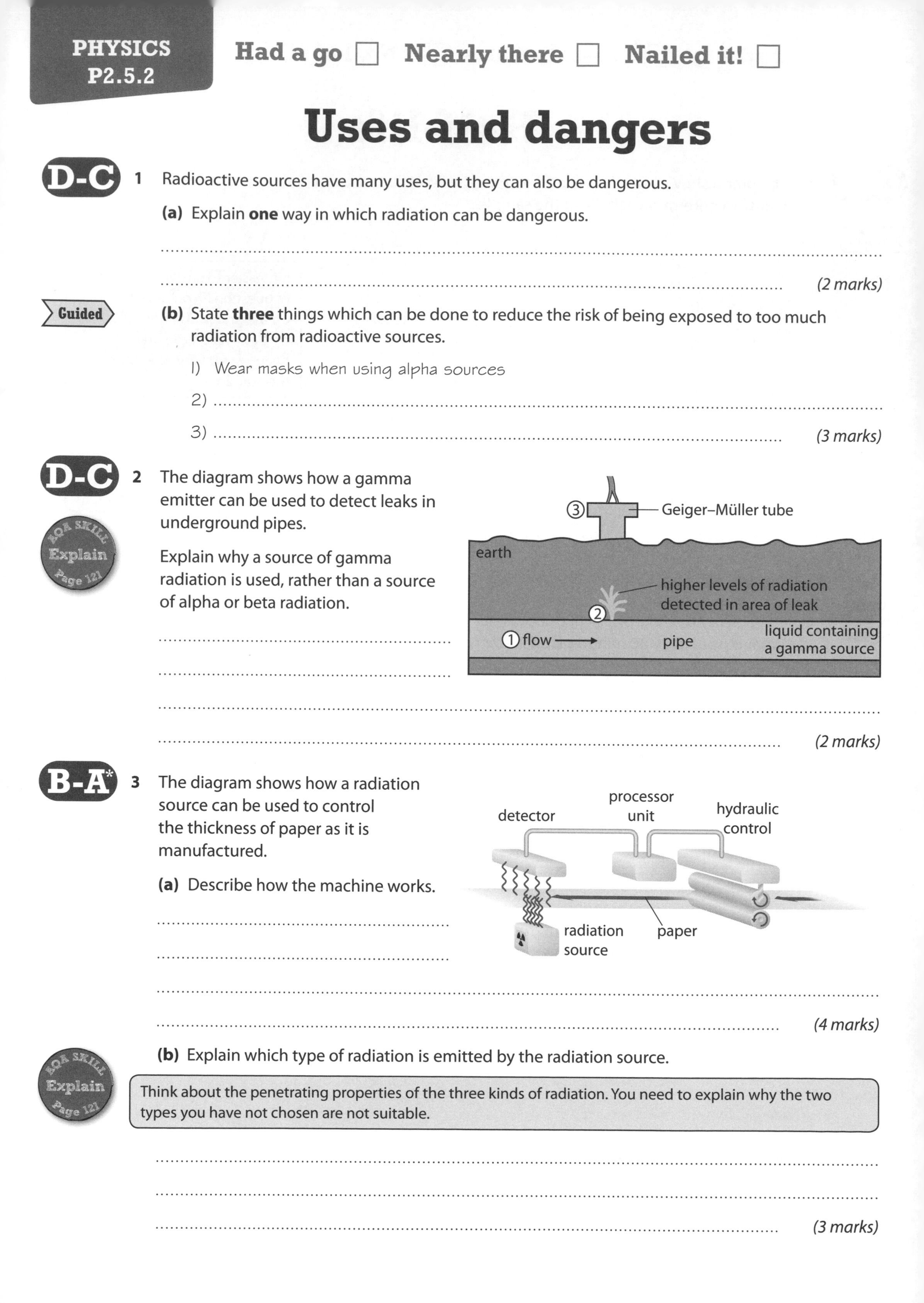

Had a go ☐ Nearly there ☐ Nailed it! ☐

Uses and dangers

D-C **1** Radioactive sources have many uses, but they can also be dangerous.

(a) Explain **one** way in which radiation can be dangerous.

....

.... *(2 marks)*

Guided **(b)** State **three** things which can be done to reduce the risk of being exposed to too much radiation from radioactive sources.

1) Wear masks when using alpha sources

2) ....

3) .... *(3 marks)*

D-C **2** The diagram shows how a gamma emitter can be used to detect leaks in underground pipes.

EQA SKILL Explain Page 121

Explain why a source of gamma radiation is used, rather than a source of alpha or beta radiation.

....

....

....

.... *(2 marks)*

B-A* **3** The diagram shows how a radiation source can be used to control the thickness of paper as it is manufactured.

(a) Describe how the machine works.

....

....

....

.... *(4 marks)*

(b) Explain which type of radiation is emitted by the radiation source.

EQA SKILL Explain Page 121

Think about the penetrating properties of the three kinds of radiation. You need to explain why the two types you have not chosen are not suitable.

....

....

.... *(3 marks)*

The nuclear model of the atom

In 1909 Marsden carried out an experiment that involved firing alpha particles at a thin sheet of gold foil.

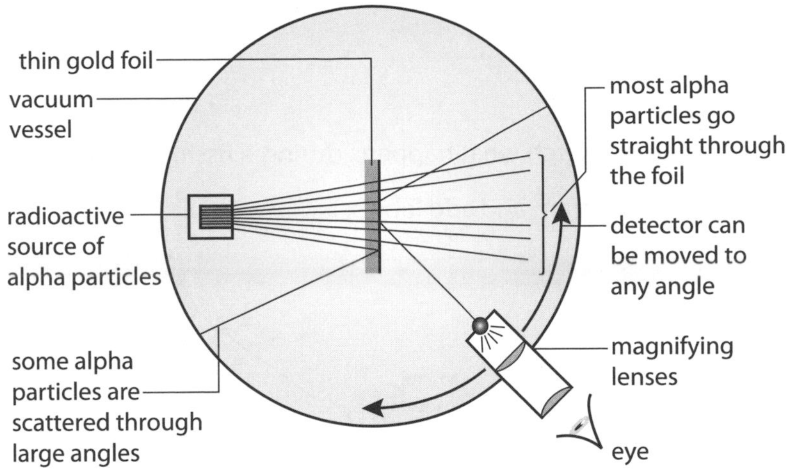

thin gold foil
vacuum vessel
radioactive source of alpha particles
some alpha particles are scattered through large angles
most alpha particles go straight through the foil
detector can be moved to any angle
magnifying lenses
eye

D-C **1** Each statement below is evidence for the structure of the nuclear atom.
Read each statement and explain what it is evidence for:

(a) 'The vast majority of alpha particles fired straight at the gold nucleus passed straight through it.'

.... *(1 mark)*

(b) 'A few alpha particles bounced back the way they came.'

.... *(1 mark)*

D-C **2** Why did the experiment with alpha particles and gold foil make scientists think again about the structure of the atom?

....

.... *(2 marks)*

B-A* **3** Compare the plum pudding and nuclear models of the atom.

Guided

> The command word here is **compare**. This means you need to list some similarities and differences.

Both models included positive and .... In the plum pudding model

....

....

.... *(3 marks)*

B-A* **4** A few of the alpha particles that Marsden fired at the gold foil bounced back the way they came.

AQA SKILL Explain Page 121

(a) Explain how this observation indicated that the positive charge was concentrated in the centre of the atom.

....

.... *(2 marks)*

(b) Explain how this observation indicated that the mass of the atom was concentrated in the centre.

....

.... *(2 marks)*

Nuclear fission

D-C **1** Name **two** elements that can undergo nuclear fission.

Guided

Uranium-235 and ....

(2 marks)

D-C **2** A student started to sketch what happens during a fission reaction.

(a) Complete the diagram and add labels.

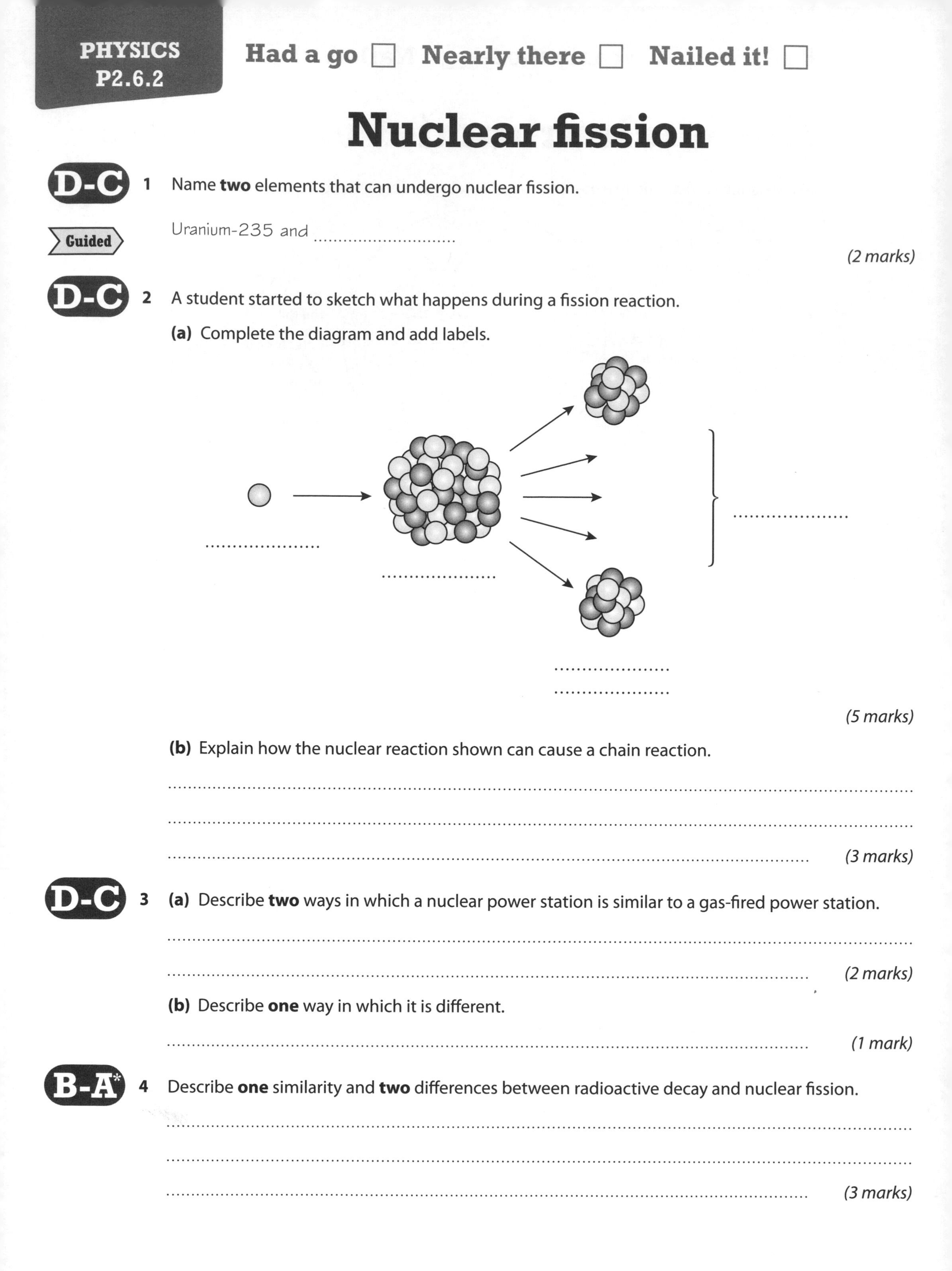

(5 marks)

(b) Explain how the nuclear reaction shown can cause a chain reaction.

....

....

.... *(3 marks)*

D-C **3** **(a)** Describe **two** ways in which a nuclear power station is similar to a gas-fired power station.

....

.... *(2 marks)*

(b) Describe **one** way in which it is different.

.... *(1 mark)*

B-A* **4** Describe **one** similarity and **two** differences between radioactive decay and nuclear fission.

....

....

.... *(3 marks)*

Nuclear fusion

D-C

1 Describe **one** similarity and **one** difference between nuclear fission and nuclear fusion.

Guided

Both produce new elements. In nuclear fission this is done by ....

.... but in fusion it happens when

....

(3 marks)

D-C

2 **(a)** Which element was formed in the Big Bang? .... *(1 mark)*

(b) The box shows information about seven different elements.

12 C carbon 6	56 Fe iron 26	4 He helium 2	207 Pb lead 82	232 Th thorium 90	238 U uranium 92	89 Y yttrium 39

(i) Write down the symbols for the elements that are formed inside stars.

.... *(2 marks)*

(ii) When are the other elements in the list formed?

.... *(1 mark)*

B-A*

3 The Earth contains most of the elements in the Periodic Table. Explain how these elements got into the dust from which the Earth was formed.

....

....

.... *(3 marks)*

B-A*

4 Explain why fission reactions and not fusion reactions are used to generate electricity.

AQA SKILL Explain Page 121

EXAM ALERT

Make sure you are clear on the difference between nuclear fission and nuclear fusion.

Students have struggled with questions like this in recent exams – **be prepared!**

....

....

....

.... *(4 marks)*

Had a go ☐ Nearly there ☐ Nailed it! ☐

The lifecycle of stars

1 The chart shows what happens to two different stars towards the end of their life cycle. Complete the missing labels.

Guided

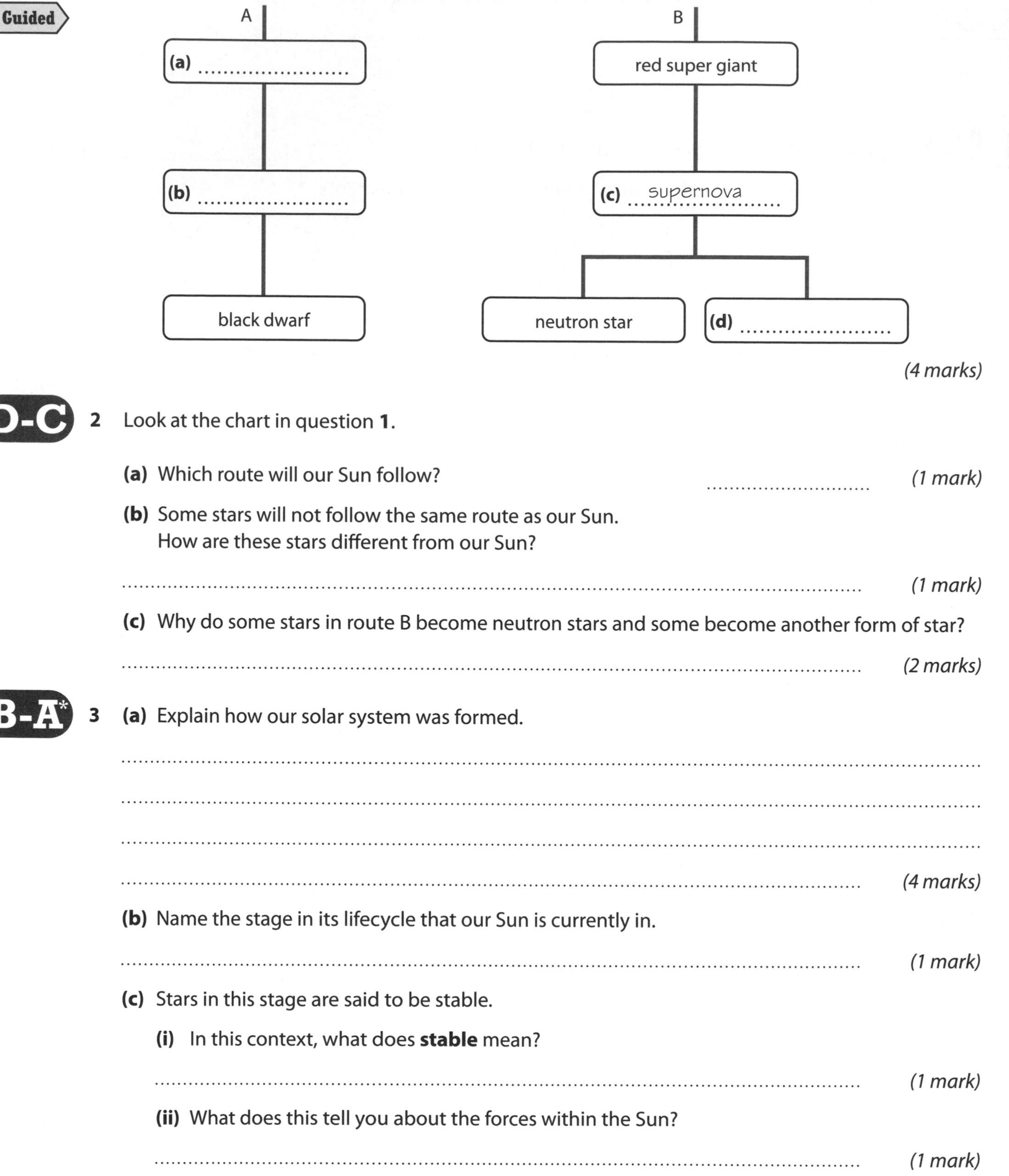

(4 marks)

D-C **2** Look at the chart in question **1**.

(a) Which route will our Sun follow? .... *(1 mark)*

(b) Some stars will not follow the same route as our Sun.
How are these stars different from our Sun?

.... *(1 mark)*

(c) Why do some stars in route B become neutron stars and some become another form of star?

.... *(2 marks)*

B-A* **3** **(a)** Explain how our solar system was formed.

....

....

....

.... *(4 marks)*

(b) Name the stage in its lifecycle that our Sun is currently in.

.... *(1 mark)*

(c) Stars in this stage are said to be stable.

(i) In this context, what does **stable** mean?

.... *(1 mark)*

(ii) What does this tell you about the forces within the Sun?

.... *(1 mark)*

Physics six mark question 3

The table shows some different radioactive isotopes.

Isotope	Radiation emitted	Half-life
phosphorus-32	beta	14 days
technetium-99m	gamma	6 hours
americium-241	alpha	432 years
barium-137m	gamma	2.5 minutes
polonium-210	alpha	138 days

Explain which isotope would be best for a medical tracer, and which would be best to use in a smoke alarm.

You will be more successful in six mark questions if you plan your answer before you start writing.

You could include information on the following:

- the properties of the different types of radiation (penetration and ionising ability)
- what **half-life** means
- what a medical tracer is used for
- the properties needed by a radioisotope used as a medical tracer
- how a smoke alarm works
- the properties needed by a radioisotope used in smoke alarms.

(6 marks)

PRACTICE PAPER

Additional Science Biology B2 practice paper

Time allowed: 60 minutes

This practice exam paper has been written to help you practise what you have learned and may not be representative of a real exam paper.

1 The diagram shows a typical plant cell.

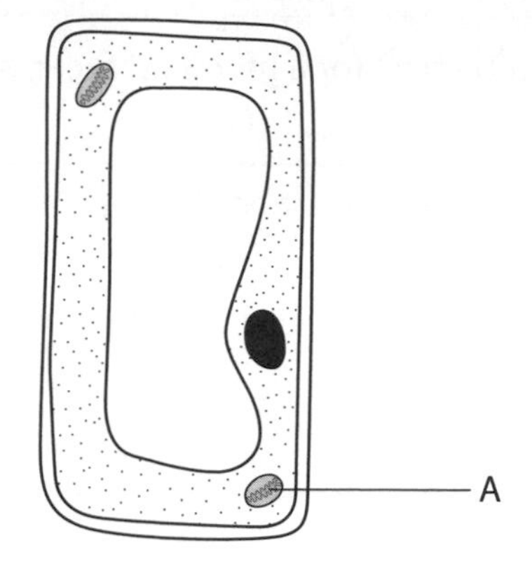

(a) (i) Draw a ring around the correct answer in the box to complete the sentence.

The part of this plant cell labelled A is the [chloroplast / cytoplasm / nucleus].

(1 mark)

(ii) Describe the function of the part of the cell labelled A.

....

....

.... *(2 marks)*

(b) Other than part A, label **one** part of this cell that is not found in animal cells.

(1 mark)

(c) Suggest **one** way in which this cell would be adapted if it were in the mesophyll of a plant.

.... *(1 mark)*

2 In this question you will be assessed on using good English, organising information clearly and using specialist terms where appropriate.

A group of students noticed that different plant species were found at different distances from a main road.

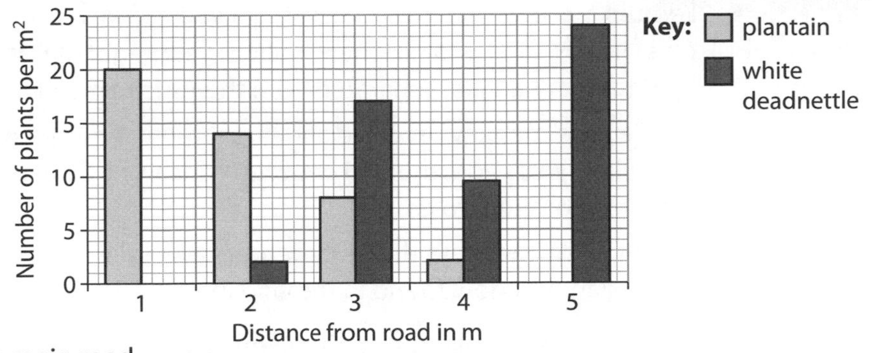

They did an investigation to look at how the distribution of two of these plant species changed with distance away from the road. They looked at plantain and white deadnettle plants.

At the end of their investigation, they were able to plot the graph above.

Describe how the students would have collected valid and reproducible data to plot in this graph.

....

....

....

....

....

....

.... *(6 marks)*

3 A runner is training for a long-distance race. During her training session, she runs at a constant speed on a running machine for 15 minutes. A trainer measures the runner's heart rate every 3 minutes during the training session.

Here are the data that the trainer collected.

Time in minutes	Heart rate in beats per minute
0	60
3	95
6	100
9	105
12	110
15	110

(a) Describe the pattern shown by the data.

....

.... *(2 marks)*

(b) During the 15 minute training session, the heart rate of the runner increases by 83%. Why is it important that this change in heart rate occurs as the runner trains?

....

....

.... *(3 marks)*

(c) The trainer explained to the runner that lactic acid could build up in her muscle cells during the training session.

(i) Name the process that produces lactic acid.

.... *(1 mark)*

(ii) Describe the effect of lactic acid on the muscle cells of the runner.

....

.... *(2 marks)*

4 Lemurs are small animals that first evolved in Africa, and began to cross into Madagascar about 60 million years ago. At this time, Madagascar had just separated from the African mainland.

By the time monkeys evolved and appeared in Africa, around 20 million years ago, the gap between Africa and Madagascar was too large for them to cross.

Africa

Madagascar

In Africa, the monkeys drove the lemurs toward extinction.
Lemurs are not now found in the wild in Africa, but are found only in Madagascar.

(a) Between 60 million and 20 million years ago, the lemurs in Madagascar evolved in a different way to the lemurs that remained in Africa. What term is used to describe the way in which new lemur species arose in Madagascar?

.... *(1 mark)*

(b) Explain whether lemurs can be described as extinct.

....

.... *(2 marks)*

(c) Describe how monkeys might have driven lemurs towards extinction in Africa.

....

.... *(2 marks)*

(d) Scientists rely on fossils to work out how lemurs and other primates evolve.
Give **one** problem that scientists may experience when using fossils for this process.

.... *(1 mark)*

5 Sexual reproduction takes place when gametes from parent organisms join at fertilisation.

(a) Name the **two** different types of gamete formed by parent organisms.

.... *(1 mark)*

(b) Describe the cell division process that results in the formation of new gametes.

....

....

.... *(3 marks)*

(c) The 2011 census of the United Kingdom showed that 49.1% of the population was male and 50.9% was female. Males have the chromosomes XY and females have the chromosomes XX.

(i) Use a genetic diagram to explain why the population of the United Kingdom has approximately equal numbers of males and females.

....

.... *(3 marks)*

(ii) Suggest **one** reason why the ratio of males to females in the United Kingdom is not exactly 50 : 50.

.... *(1 mark)*

6 The diagram shows two mice. One is a normal mouse with brown fur; the other is an albino mouse.

normal mouse albino mouse

Animals that are albinos do not have coloured pigment in their skin or hair, so this mouse has white fur.

Albinism is an inherited condition controlled by a single gene with two alleles. The condition is homozygous recessive.

(a) What is meant by the phrase **homozygous recessive**?

....

.... *(2 marks)*

(b) Two normal mice with brown fur are mated together. They produced some offspring with normal (brown) fur and some offspring with white (albino) fur.

Draw a Punnett square to show how two parent mice with brown fur can produce offspring that are albino mice. Use **A** to represent the normal allele and **a** to represent the allele for albinism.

(4 marks)

(c) Both albino and brown mice exist in the wild. Biologically, both mice are of the same species. Explain what might happen if both types of mice were released on two different islands with different climates, one snowy and one warm.

....

....

.... *(3 marks)*

7 Enzymes are biological catalysts. Many enzymes occur naturally in the human body, for use in processes such as digestion. In addition to these natural enzymes, scientists have developed enzymes for use in industry.

One enzyme frequently used in industry is isomerase.

(a) Explain why isomerase is a useful enzyme in the food industry.

....

....

.... *(3 marks)*

(b) Some scientists investigate how temperature affects the action of isomerase. They collect data at different temperatures and plot this graph.

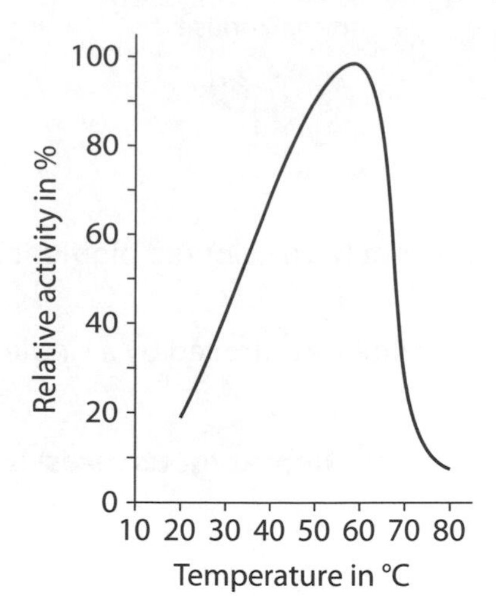

Explain the shape of this graph.

....

....

.... *(3 marks)*

8 The following story appeared in newspapers in April 2011.

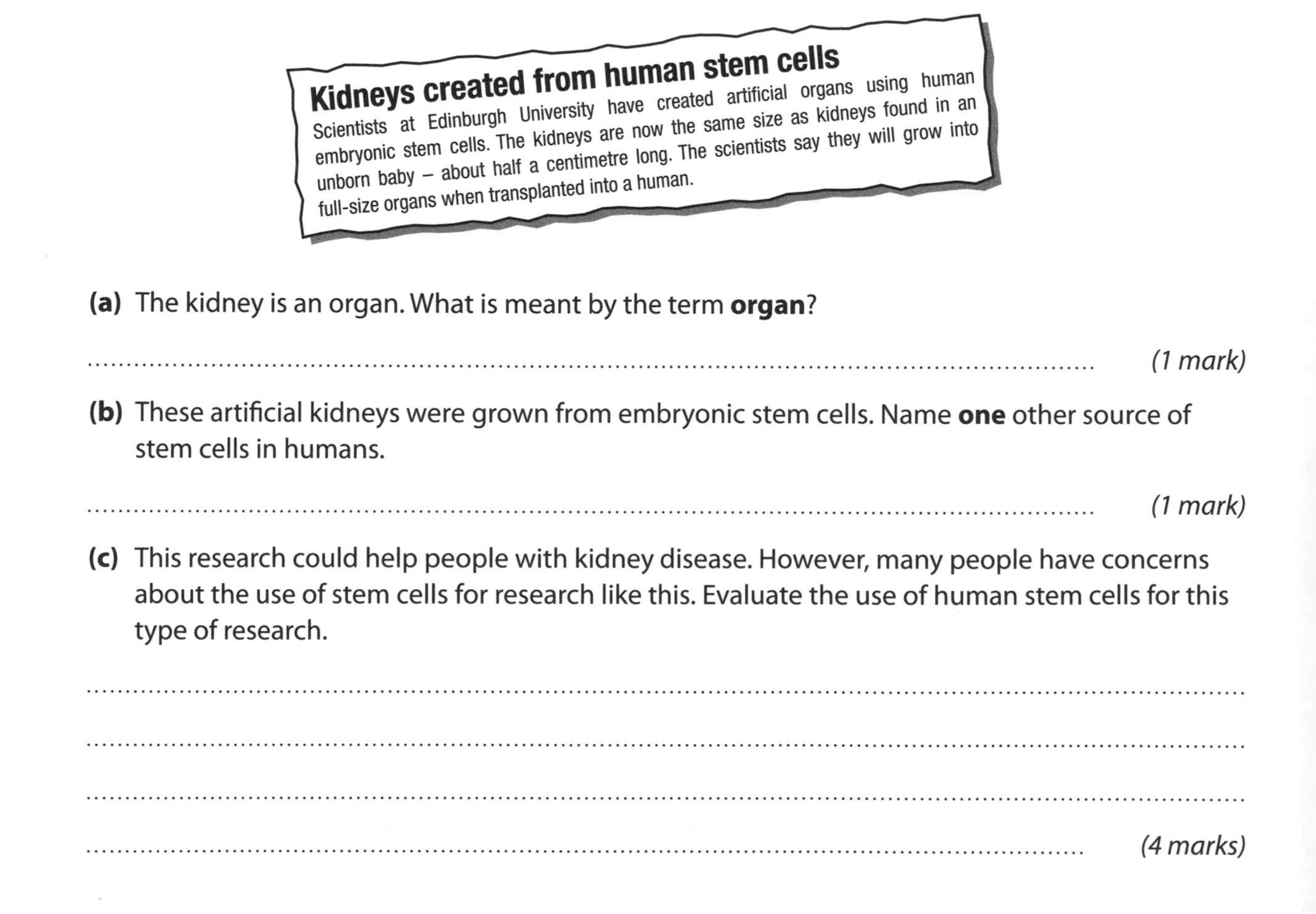

Kidneys created from human stem cells

Scientists at Edinburgh University have created artificial organs using human embryonic stem cells. The kidneys are now the same size as kidneys found in an unborn baby – about half a centimetre long. The scientists say they will grow into full-size organs when transplanted into a human.

(a) The kidney is an organ. What is meant by the term **organ**?

.... *(1 mark)*

(b) These artificial kidneys were grown from embryonic stem cells. Name **one** other source of stem cells in humans.

.... *(1 mark)*

(c) This research could help people with kidney disease. However, many people have concerns about the use of stem cells for research like this. Evaluate the use of human stem cells for this type of research.

....

....

....

.... *(4 marks)*

9 In order to grow, plants need to produce glucose by photosynthesis. This glucose can be used to make proteins and to make new cells.

(a) Other than growth, describe **one** way in which plants use the products of photosynthesis.

....

.... *(2 marks)*

(b) A farmer is growing tomatoes in a greenhouse. She wants to see how fast the tomato plants photosynthesise during the day.

She maintains her greenhouse at a constant temperature of 30 °C. Every hour between 06:00 and 18:00 she measures the light intensity in the greenhouse and the rate of photosynthesis of the tomato plants. She plots her results on this graph.

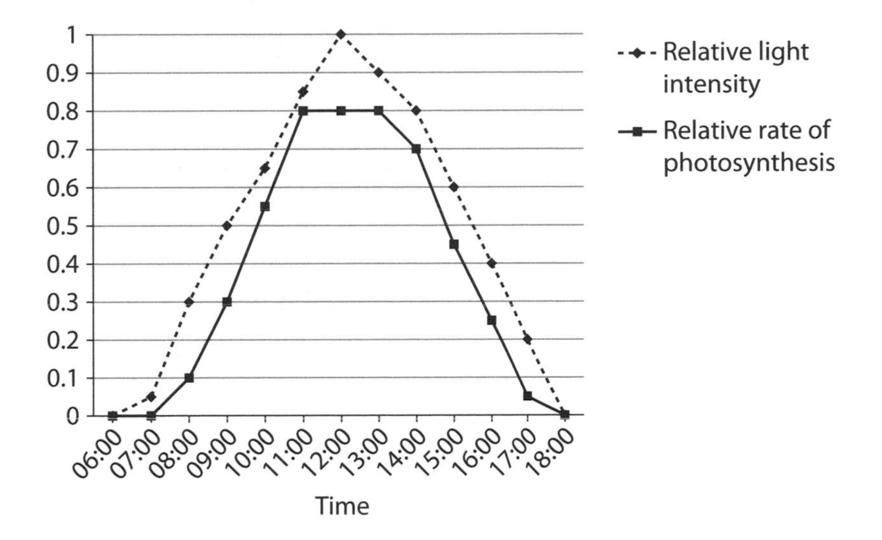

Explain the shape of the graph for the rate of photosynthesis.

....

....

....

.... *(4 marks)*

Additional Science Chemistry C2 practice paper

Time allowed: 60 minutes

This practice exam paper has been written to help you practise what you have learned and may not be representative of a real exam paper.

1 The diagram below shows the structure of a compound.

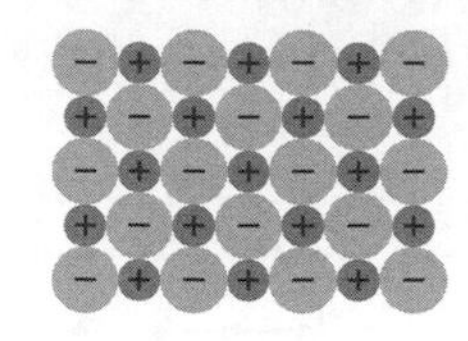

(a) Name the type of bonding and structure shown in the diagram.

.... *(2 marks)*

(b) Describe the forces of attraction involved in this structure.

....

.... *(2 marks)*

(c) Explain whether this compound will conduct electricity when it is a solid.

....

.... *(2 marks)*

2 **(a)** Calcium can react with oxygen to form a compound. The electron arrangements of calcium and oxygen are shown below.

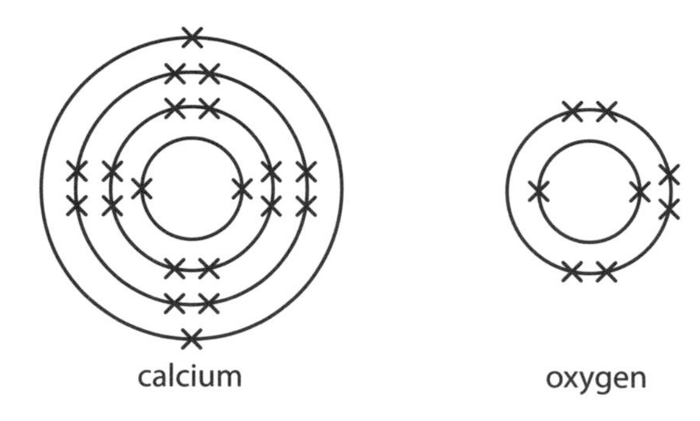

(i) Complete the diagrams showing the ions formed by calcium and oxygen when they react.

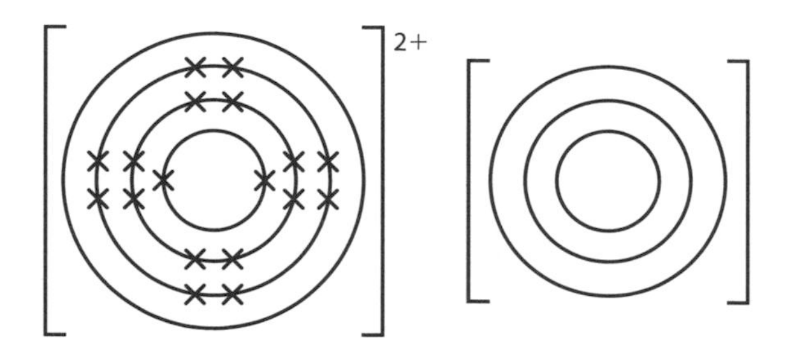

(2 marks)

(ii) Complete the balanced equation for the reaction between calcium and oxygen:

.... + ... $\rightarrow$ 2CaO

(2 marks)

(b) Calcium also forms a compound with chlorine

(i) Write the formula for calcium chloride. .... *(1 mark)*

(ii) What is the charge on the calcium ion in calcium chloride? .... *(1 mark)*

(iii) Name an element in the first 20 of the Periodic Table with similar properties to chlorine. .... *(1 mark)*

(c) Which element has atoms with the same electron arrangement as a chloride ion? .... *(1 mark)*

3 A new form of carbon, a fullerene, was discovered in 1985. The first molecules to be isolated had the formula C_{60}, and several other molecules have been discovered since then.

Scientists have been researching and developing uses for fullerenes. One area of research uses these nanoparticles to deliver drugs to different parts of the body. The cage-like molecules contain the drug and are absorbed as they can easily pass through the walls of human cells.

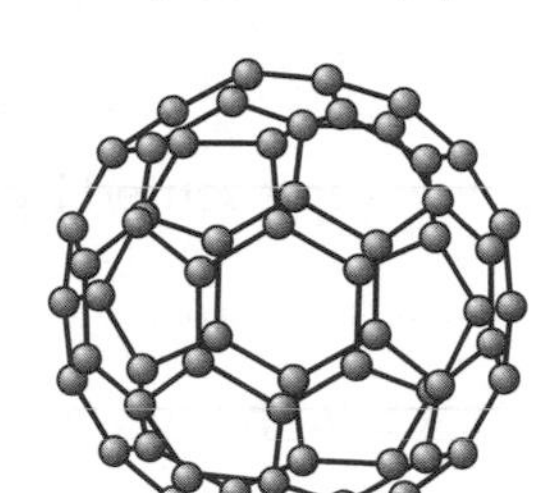

(a) Why can nanoparticles pass easily through cell walls?

.... *(1 mark)*

(b) Suggest **one** possible problem with the use of nanoparticles in medicines.

.... *(1 mark)*

(c) Name a different form of carbon and describe how its structure is different from fullerene.

....

.... *(2 marks)*

(d) Nanoparticles are also used as catalysts in the chemical industry. For example, nanoparticles of gold are used in the manufacture of polymers like poly(vinyl acetate).

Explain why using an expensive catalyst like gold can be cost-effective in the manufacture of polymers like poly(vinyl acetate).

....

.... *(2 marks)*

4 The water in a swimming pool is chlorinated to kill bacteria. The chlorination is carried out in a chlorinator cell in which electrolysis of sodium chloride solution produces chlorine gas that dissolves in the water.

(a) Complete the labels on the diagram of the chlorinator cell, which is used to produce chlorine gas.

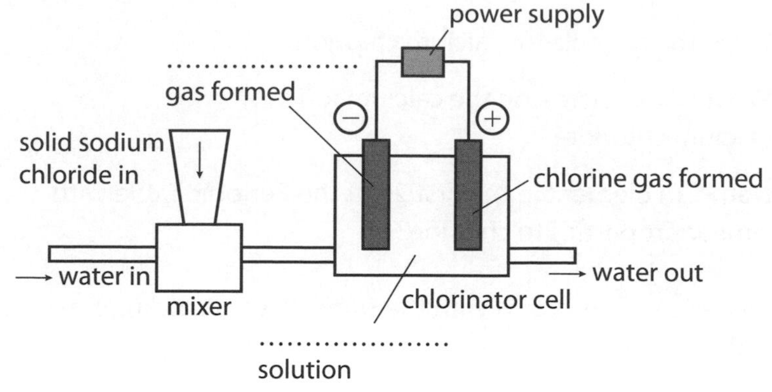

(2 marks)

(b) (i) Complete the balanced half equation for the formation of chlorine gas at the electrode during electrolysis.

.... $\rightarrow Cl_2(g)$ + ...

(2 marks)

(ii) Name the type of reaction represented by this half equation.

.... *(1 mark)*

(c) Some of the chlorine gas reacts with the water as shown below.

$Cl_2(g) + H_2O(l) \rightarrow 2H^+(aq) + Cl^-(aq) + OCl^-(aq)$

Explain why this reaction causes the pH of the solution to decrease.

....

.... *(2 marks)*

5 Sulfuric acid is widely used throughout the chemical industry. It is manufactured by the contact process, starting with sulfur. The process is summarised below.

Step 1: Solid sulfur is burned with the oxygen in air to produce sulfur dioxide gas.

Step 2: The sulfur dioxide gas is passed with more air over a catalyst of vanadium pentoxide to produce sulfur trioxide.

$2SO_2(g) + O_2(g) \rightleftharpoons 2SO_3(g)$

Step 3: The sulfur trioxide is dissolved in sulfuric acid to make oleum.

$SO_3(g) + H_2SO_4(l) \rightarrow H_2S_2O_7(l)$

Step 4: The oleum is dissolved in water to make more sulfuric acid.

$H_2S_2O_7\,(l) + H_2O(l) \rightarrow 2H_2SO_4(l)$

(a) Complete the balanced equation, including state symbols, for Step 1.

S(s) + ...(.........) $\rightarrow SO_2(g)$

(1 mark)

(b) What does the sign $\rightleftharpoons$ in the equation in Step 2 tell you about the reaction?

.... *(1 mark)*

(c) The graph shows how temperature affects the percentage conversion of SO_2 to SO_3 in Step 2.

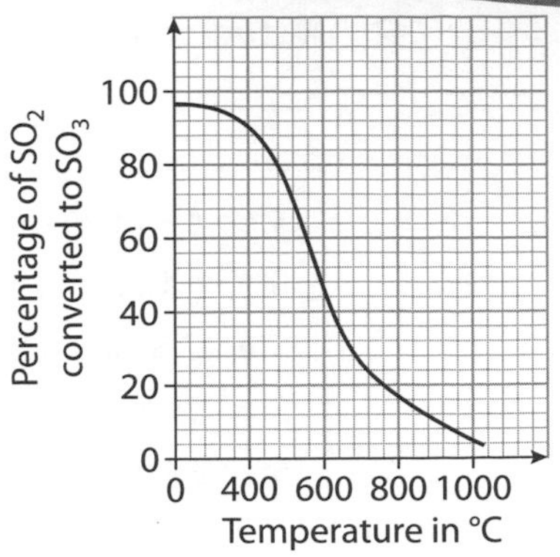

(i) State how the percentage conversion to SO_3 depends on temperature.

....

....

(1 mark)

(ii) Calculations show that 8 tonnes of sulfur dioxide should produce 10 tonnes of sulfur trioxide.

Use the graph to find the percentage conversion at 500 °C and calculate the actual yield of sulfur trioxide, from 8 tonnes of sulfur dioxide, at this temperature.

....

....

.... *(3 marks)*

(d) Using the equation from step 3, calculate the mass of H_2SO_4 ($M_r = 98$) needed to react with 400 g of SO_3 ($M_r = 80$).

....

....

....

mass of H_2SO_4 = ... g *(2 marks)*

6 In this question you will be assessed on using good English, organising information clearly and using specialist terms where appropriate.

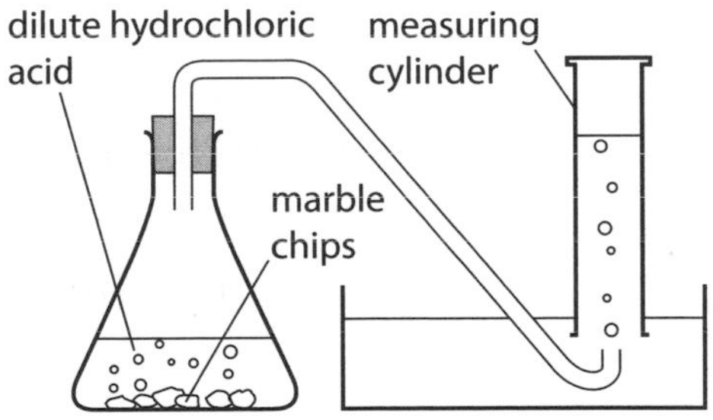

The reaction between marble chips and hydrochloric acid produces carbon dioxide gas.

The speed of this reaction depends on a number of factors, including temperature, concentration and particle size. Using ideas about how chemical reactions occur, explain how these three factors affect the rate of reaction between marble chips and hydrochloric acid.

....

....

....

....

....

....

....

....

....

.... *(6 marks)*

7 A student carried out an experiment, using the workcard below, to make cobalt chloride crystals.

Workcard – Making cobalt chloride crystals
Step A: Put 25 cm^3 of dilute hydrochloric acid into a 100 cm^3 beaker.
Step B: Use a spatula to add cobalt carbonate powder to the beaker.
Step C: Stop adding cobalt carbonate when ....
Step D: Filter the mixture to remove unreacted cobalt carbonate.
Step E: Leave the solution in a warm place to form crystals of cobalt chloride.

(a) Complete the instruction in step C.

.... *(1 mark)*

(b) Explain why filtering removes the unreacted cobalt carbonate.

....

....

.... *(2 marks)*

(c) Complete the balanced equation for this reaction, including state symbols.

$CoCO_3(s) +$ .... $\rightarrow CoCl_2(aq) + H_2O(l) +$ ....

(2 marks)

(d) Suggest why cobalt chloride could not be made by adding cobalt metal to dilute hydrochloric acid.

....

.... *(1 mark)*

8 The atoms of lithium are made up of three main subatomic particles, called protons, neutrons and electrons.

(a) Describe the structure of an atom.

....

.... *(2 marks)*

(b) Two isotopes of lithium metal are ${}^{7}_{3}Li$ and ${}^{6}_{3}Li$.

(i) How are these two isotopes different to each other?

.... *(1 mark)*

(ii) Explain why these two atoms will react in the same way even though they are not identical.

....

.... *(2 marks)*

(c) Explain how lithium metal is able to conduct electricity in the solid state.

....

.... *(2 marks)*

9 Toothpaste contains a number of different substances and each of the substances has a particular job to do.

Type of substance	Example	Job
abrasive	silicon dioxide	to scrape off plaque and stains
detergent	sodium lauryl sulfates	to produce foam in the mouth
fluoride	sodium fluoride	to protect the tooth enamel
base	sodium hydroxide	to neutralise acids
flavouring	sorbitol	to make toothpaste taste better
colouring agent	artificial dyes	to make toothpaste look better

(a) The silicon dioxide structure is shown opposite. Explain why this structure makes it suitable for use as an abrasive.

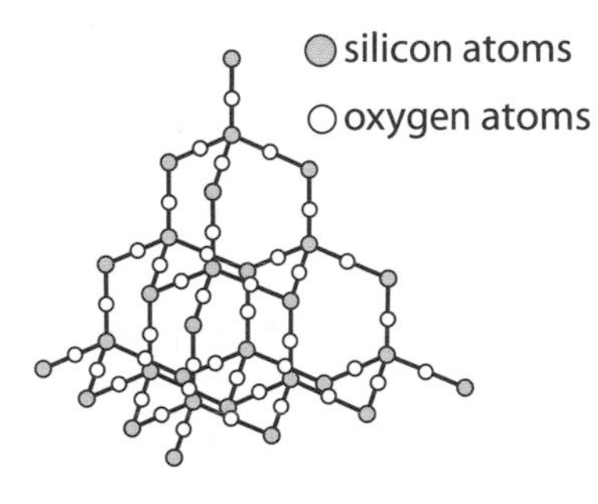

....

.... *(2 marks)*

(b) Fluorides, like sodium fluoride, have recently been added to our water supply in an effort to improve dental health. Some people objected to the addition of fluorides to their water supply.

Suggest **two** reasons why some people might be unhappy with the addition of fluoride to their water supply.

....

.... *(2 marks)*

(i) Sodium hydroxide is an alkali. What ion in acids does it react with?

.... *(1 mark)*

(ii) Complete the equation for the reaction between sodium hydroxide and nitric acid.

$NaOH + HNO_3 \rightarrow NaNO_3 +$ ....

(1 mark)

PRACTICE PAPER

Additional Science Physics P2 practice paper

Time allowed: 60 minutes

This Practice Exam Paper has been written to help you practise what you have learned and may not be representative of a real exam paper.

1 The diagram shows a painter working at the top of a ladder.

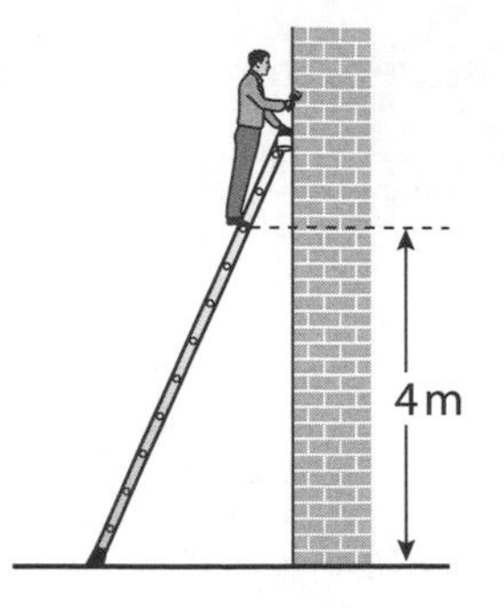

(a) The painter has a mass of 75 kg. Calculate the weight of the painter. Use a value for *g* of 10 N/kg.

Write down the equation you use, and then show clearly how you work out your answer.

....

....

Weight = ... N *(2 marks)*

(b) The painter has climbed 4 m from the ground. Calculate the work done by the painter when he climbs this distance.

Write down the equation you use, and then show clearly how you work out your answer.

....

....

Work done = ... J *(2 marks)*

(c) The pot of paint has a mass of 0.5 kg. The painter is holding it 5 m above the ground. The gravitational potential energy of the pot of paint is 25 J.

(i) The painter drops the pot of paint. What will the kinetic energy of the pot be just before it hits the ground? Explain your answer.

....

.... *(2 marks)*

(ii) Calculate the velocity of the paint pot as it hits the ground.

Write down the equation you use, and then show clearly how you work out your answer.

....

....

Velocity = ... m/s *(3 marks)*

2 A scientist is investigating the types of radiation given off by a radioactive source.

(a) Suggest **two** safety precautions the scientist should follow while carrying out the experiments.

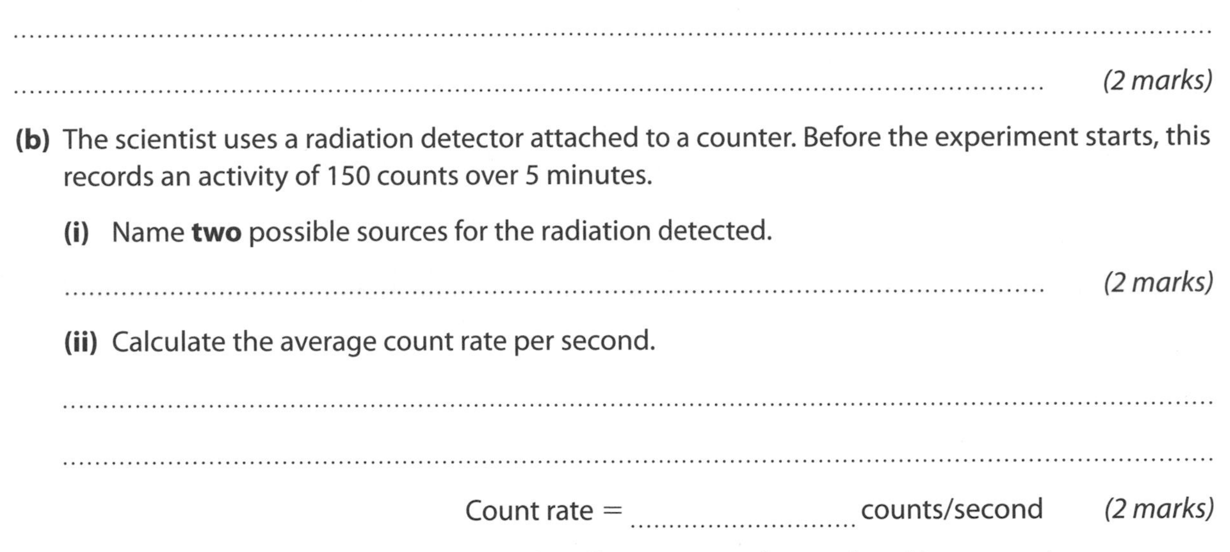

....

.... *(2 marks)*

(b) The scientist uses a radiation detector attached to a counter. Before the experiment starts, this records an activity of 150 counts over 5 minutes.

(i) Name **two** possible sources for the radiation detected.

.... *(2 marks)*

(ii) Calculate the average count rate per second.

....

....

Count rate = ... counts/second *(2 marks)*

(c) The diagram shows the apparatus used. Different materials are placed between the source and the detector.

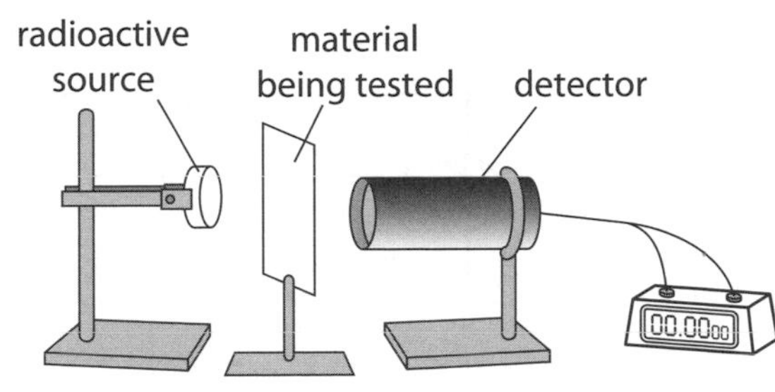

The table shows the results of this experiment.

Material	Counts per second
none	700
single sheet of paper	695
aluminium sheet 1 mm thick	476
aluminium sheet 3 mm thick	154

Which type or types of radiation is the source emitting? Explain your answer.

....

....

....

.... *(4 marks)*

(d) The activity of the source was recorded over several hours. The table shows the corrected counts recorded.

Time in hours	Counts per second
0	700
1	445
2	276
3	175
4	110
5	70

(i) Plot these results on the axes below. Join the points with a smooth curve.

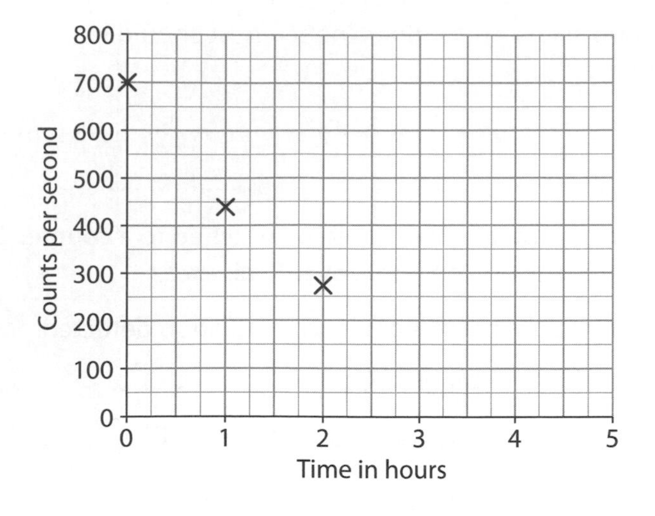

(2 marks)

(ii) Use your graph to determine the half-life of the source.

.... *(2 marks)*

3 The diagram below shows three identical bulbs connected to a 12 V battery.

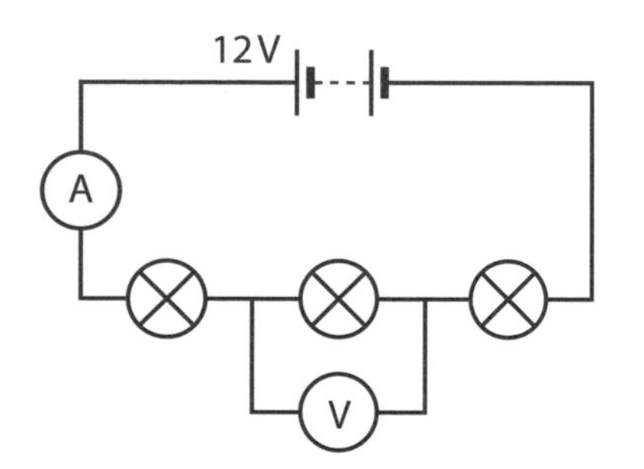

(a) What is the size of the potential difference across each bulb?

.... *(1 mark)*

(b) The current flowing through each bulb is 1.6 A. Calculate the resistance of one bulb.

Write down the equation you use, and then show clearly how you work out your answer.

....

....

Resistance = ... Ω *(2 marks)*

(c) Calculate the power of one bulb in the circuit shown.

Write down the equation you use, and then show clearly how you work out your answer.

....

....

Power = ... W *(2 marks)*

(d) A different circuit is set up using different bulbs. These bulbs have a resistance of 4 Ω.

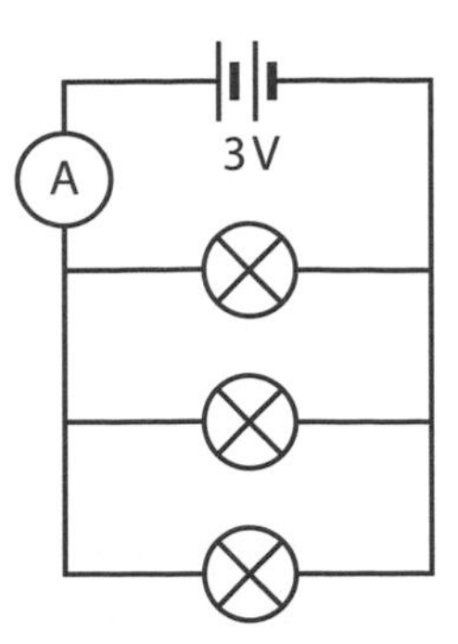

(i) What is the size of the voltage across each bulb?

.... *(1 mark)*

(ii) Calculate the ammeter reading in this circuit.

Write down the equation you use, and then show clearly how you work out your answer.

....

....

Ammeter reading = ... A *(3 marks)*

4 A car has a mass of 1500 kg and is travelling at 12 m/s. The driver sees a road sign, and applies the brakes a few seconds later. The car slows to a stop. The graph shows the velocity of the car during this time.

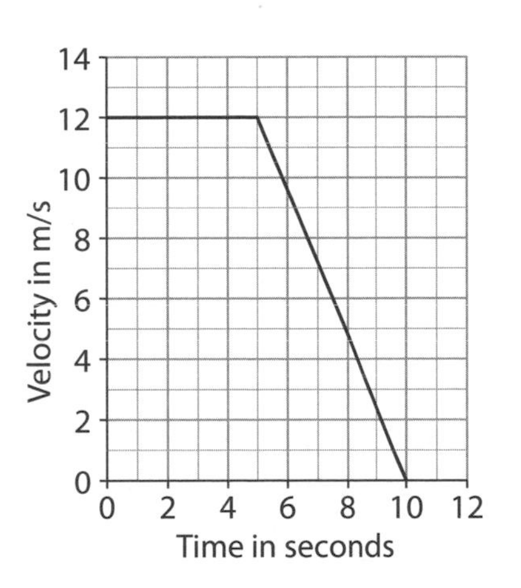

(a) Determine how far the car travelled while the brakes were applied.

Write down the equation you use, and then show clearly how you work out your answer.

....

....

Distance = ... m *(2 marks)*

(b) Calculate the kinetic energy of the car when it was moving at 12 m/s.

Write down the equation you use, and then show clearly how you work out your answer.

....

....

....

....

Kinetic energy = ... J *(2 marks)*

(c) Calculate the force applied by the brakes. Write down the equation you use, and then show clearly how you work out your answer.

....

....

....

Force = ... N *(3 marks)*

(d) In this question you will be assessed on using good English, organising information clearly and using specialist terms where appropriate.

The speed limit for cars travelling on a motorway is 70 mph. However, if the car is towing a caravan, the speed limit is 60 mph.

Explain why there are two different speed limits. Use ideas about forces and energy in your answer.

....

....

....

....

....

....

....

....

....

....

.... *(6 marks)*

5 An astronaut at the International Space Station has a mass of 65 kg and a tool bag that has a mass of 13 kg. She throws the tool bag and it leaves her hand with a velocity of 2 m/s.

(a) Calculate the momentum of the tool bag. Write down the equation you use and then show clearly how you work out your answer. Give the units for your answer.

....

....

....

Momentum = ... *(3 marks)*

(b) Calculate the velocity of the astronaut.

....

....

....

Velocity = ... m/s *(3 marks)*

6 A baby bouncer consists of a harness for the baby that hangs in a doorway. There is a spring that allows the baby to bounce up and down.

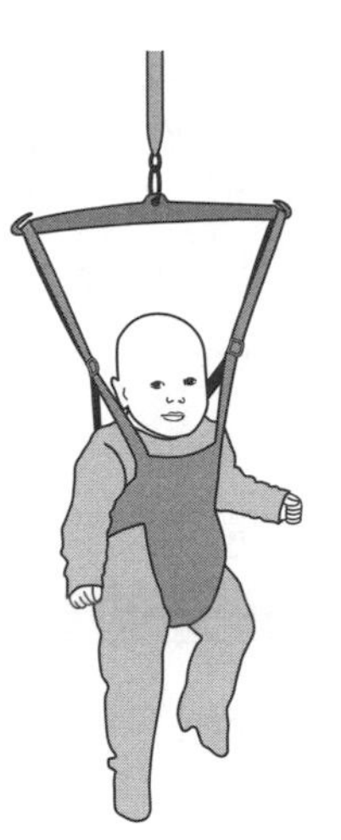

A parent wants to allow their baby to bounce up and down further. The bouncer is currently fitted with spring B. Explain which spring the parent should fit to the baby bouncer.

Spring	*k* in N/m
A	450
B	350
C	230

....

....

.... *(3 marks)*

7 Bismuth-214 is a radioactive isotope. It decays by beta decay to form an isotope of polonium. The polonium decays to form lead-210.

(a) Complete the missing numbers in this equation.

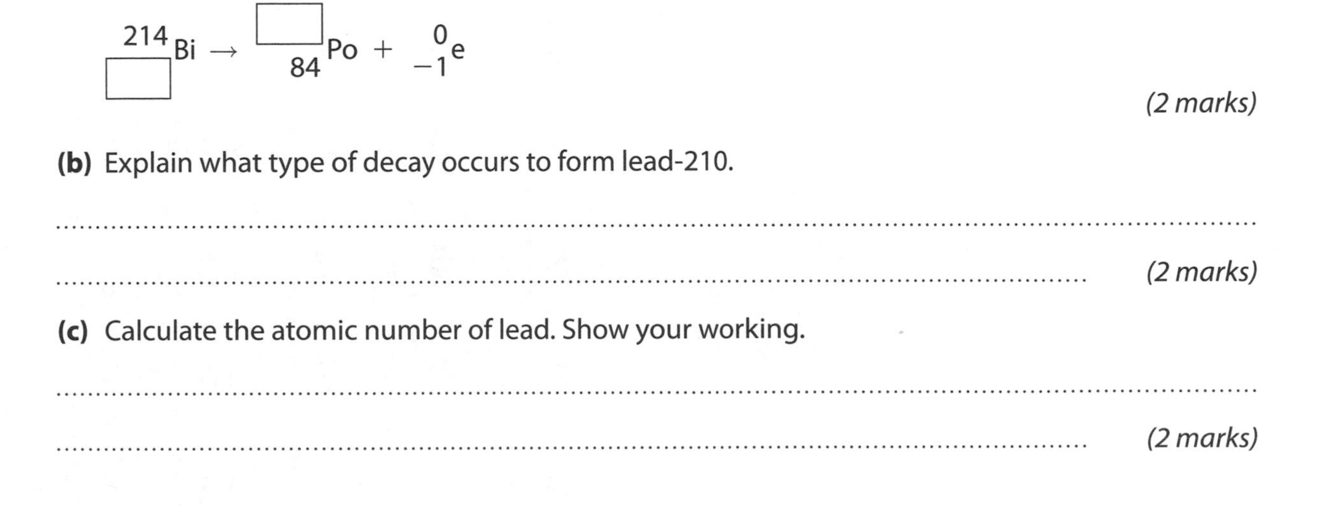

$$^{214}_{\square}\text{Bi} \rightarrow \,^{\square}_{84}\text{Po} + \,^{0}_{-1}\text{e}$$

(2 marks)

(b) Explain what type of decay occurs to form lead-210.

....

.... *(2 marks)*

(c) Calculate the atomic number of lead. Show your working.

....

.... *(2 marks)*

Key

relative atomic mass **atomic symbol** name atomic (proton) number

1	2											3	4	5	6	7	0
						1 **H** hydrogen 1											4 **He** helium 2
7 **Li** lithium 3	9 **Be** beryllium 4											11 **B** boron 5	12 **C** carbon 6	14 **N** nitrogen 7	16 **O** oxygen 8	19 **F** fluorine 9	20 **Ne** neon 10
23 **Na** sodium 11	24 **Mg** magnesium 12											27 **Al** aluminium 13	28 **Si** silicon 14	31 **P** phosphorus 15	32 **S** sulfur 16	35.5 **Cl** chlorine 17	40 **Ar** argon 18
39 **K** potassium 19	40 **Ca** calcium 20	45 **Sc** scandium 21	48 **Ti** titanium 22	51 **V** vanadium 23	52 **Cr** chromium 24	55 **Mn** manganese 25	56 **Fe** iron 26	59 **Co** cobalt 27	59 **Ni** nickel 28	63.5 **Cu** copper 29	65 **Zn** zinc 30	70 **Ga** gallium 31	73 **Ge** germanium 32	75 **As** arsenic 33	79 **Se** selenium 34	80 **Br** bromine 35	84 **Kr** krypton 36
85 **Rb** rubidium 37	88 **Sr** strontium 38	89 **Y** yttrium 39	91 **Zr** zirconium 40	93 **Nb** niobium 41	96 **Mo** molybdenum 42	99 **Tc** technetium 43	101 **Ru** ruthenium 44	103 **Rh** rhodium 45	106 **Pd** palladium 46	108 **Ag** silver 47	112 **Cd** cadmium 48	115 **In** indium 49	119 **Sn** tin 50	122 **Sb** antimony 51	128 **Te** tellurium 52	127 **I** iodine 53	131 **Xe** xenon 54
133 **Cs** caesium 55	137 **Ba** barium 56	139 **La** lanthanum 57	178 **Hf** hafnium 72	181 **Ta** tantalum 73	184 **W** tungsten 74	186 **Re** rhenium 75	190 **Os** osmium 76	192 **Ir** iridium 77	195 **Pt** platinum 78	197 **Au** gold 79	201 **Hg** mercury 80	204 **Tl** thallium 81	207 **Pb** lead 82	209 **Bi** bismuth 83	210 **Po** polonium 84	211 **At** astatine 85	222 **Rn** radon 86
223 **Fr** francium 87	226 **Ra** radium 88	227 **Ac** actinium 89	261 **Rf** rutherfordium 104	262 **Db** dubnium 105	266 **Sg** seaborgium 106	264 **Bh** bohrium 107	277 **Hs** hassium 108	268 **Mt** meitnerium 109	271 **Ds** darmstadtium 110	272 **Rg** roentgenium 111							

The lanthanides (atomic numbers 58–71) and the actinides (atomic numbers 90–103) have been omitted.
Elements with atomic numbers 112–118 have been reported but not fully authenticated.
Relative atomic masses for Cu and Cl have not been rounded to the nearest whole number.

Chemistry data sheet

Reactivity series of metals

potassium	most reactive ↑
sodium	
calcium	
magnesium	
aluminium	
carbon	
zinc	
iron	
tin	
lead	
hydrogen	
copper	
silver	
gold	
platinum	least reactive ↓

Elements in italics, though non-metals, have been included for comparison.

Formulae of some common ions

Positive ions

Name	Formula
hydrogen	H^{+}
sodium	Na^{+}
silver	Ag^{+}
potassium	K^{+}
lithium	Li^{+}
ammonium	NH_4^{+}
barium	Ba^{2+}
calcium	Ca^{2+}
copper(II)	Cu^{2+}
magnesium	Mg^{2+}
zinc	Zn^{2+}
lead	Pb^{2+}
iron(II)	Fe^{2+}
iron(III)	Fe^{3+}
aluminium	Al^{3+}

Negative ions

Name	Formula
chloride	Cl^{-}
bromide	Br^{-}
fluoride	F^{-}
iodide	I^{-}
hydroxide	OH^{-}
nitrate	NO_3^{-}
oxide	O^{2-}
sulfide	S^{2-}
sulfate	SO_4^{2-}
carbonate	CO_3^{2-}

Physics equations sheet

$a = \frac{F}{m}$ or $F = m \times a$	F	resultant force
	m	mass
	a	acceleration
$a = \frac{v - u}{t}$	a	acceleration
	v	final velocity
	u	initial velocity
	t	time taken
$W = m \times g$	W	weight
	m	mass
	g	gravitational field strength
$F = k \times e$	F	force
	k	spring constant
	e	extension
$W = F \times d$	W	work done
	F	force applied
	d	distance moved in the direction of the force
$P = \frac{E}{t}$	P	power
	E	energy transferred
	t	time taken
$E_p = m \times g \times h$	E_p	change in gravitational potential energy
	m	mass
	g	gravitational field strength
	h	change in height
$E_k = \frac{1}{2} \times m \times v^2$	E_k	kinetic energy
	m	mass
	v	speed
$p = m \times v$	p	momentum
	m	mass
	v	velocity
$I = \frac{Q}{t}$	I	current
	Q	charge
	t	time
$V = \frac{W}{Q}$	V	potential difference
	W	work done
	Q	charge
$V = I \times R$	V	potential difference
	I	current
	R	resistance
$P = I \times V$	P	power
	I	current
	V	potential difference
$E = V \times Q$	**E**	**energy**
	V	**potential difference (Higher Tier only)**
	Q	**charge**

AQA specification skills

In your AQA exam there are certain **skills** that you sometimes need to **apply** when answering a question. Questions often contain a particular **command word** that lets you know this. On this page we explain how to spot a command word and how to apply the required skill.

Note: Watch out for our Skills sticker – this points out the questions that are particularly focused on applying skills.

Command word	Skill you are being asked to apply
AQA SKILL Apply Page 121	You might be asked to **apply** what you know about a topic to a practical situation. For example: 'Apply your knowledge of series and parallel circuits to suggest the best wiring for a set of Christmas tree lights.'
AQA SKILL Compare Page 121	**Compare** how two things are similar or different. Make sure you include both of the things you are being asked to compare. For example: 'A is bigger than B, but B is lighter than A.'
AQA SKILL Consider Page 121	You will be given some information and you will be asked to **consider** all the factors that might influence a decision. For example: 'When buying a new fridge the family would need to consider the following things …'
AQA SKILL Describe Page 121	**Describe** a process or why something happens in an accurate way. For example: 'When coal is burned the heat energy is used to turn water into steam. The steam is then used to turn a turbine …'
AQA SKILL Discuss Page 121	In some questions you might be asked to make an informed judgement about a topic. This might be something like stem cell research. You should **discuss** the topic and give your **opinion** but make sure that you back it up with information from the question or your scientific knowledge.
AQA SKILL Draw Page 121	Some questions ask you to **draw** or sketch something. It might be the electrons in an atom, a graph or a ray diagram. Make sure you take a pencil, rubber and ruler into your exam.
AQA SKILL Evaluate Page 121	This is the most important one! Most of the skill statements start with **evaluate**. You will be given information and will be expected to use that information plus anything you know from studying the material in the specification to look at evidence and come to a **conclusion**. For example, if you were asked to evaluate which of two slimming programmes was better, then you might comment like this: 'In programme A people lost weight quickly to start with but then put the weight back on by the end of the sixth month. In programme B they did not lose weight so quickly to start with, but the weight loss was slow and steady and no weight was gained back by the end of the year. I therefore think that programme B is most effective.'
AQA SKILL Explain Page 121	State what is happening and **explain** why it is happening. If a question asks you to explain then it is a good idea to try to use the word 'because' in your answer. For example: 'pH 2 is the optimal pH for enzymes in the stomach because the stomach is very acidic.'
AQA SKILL Formulae Page 121	In some chemistry questions you will be expected to write chemical **formulae** for compounds. You will be given the symbols and the ions can be found on the Data Sheet on page 119.
AQA SKILL Interpret Page 121	**Interpret** the data given to you on graphs, diagrams or in tables to help answer the question. For example: 'Use the data to show what happens when…'
AQA SKILL Predict Page 121	A question may ask you to **predict** the outcome of a genetic cross or what will be formed at different electrodes in an electrolysis experiment. If you learn the patterns of genetic crosses and the rules about electrolysis then you will be able to do this.
AQA SKILL Suggest Page 121	You will be given some information about an unfamiliar situation and asked to **suggest** an answer to a question. You will not have learned the answer – you need to **apply** your knowledge to that new situation. For example: 'I think that blue is better than green because …' or 'It might be because …'

Answers

You will find some advice next to some of the answers. This is written in italics. It is not part of the mark scheme but just gives you a little more information.

Biology answers

1. Animal and plant cells

1 **(a)** plant/alga (1) *Note that algae cells and plant cells look very similar – you are not expected to tell the differences between them!*
(b) It has a cell wall (1); it has chloroplasts (1).

2 **(a)** The mitochondrion in an animal cell is where energy release occurs (1). The energy is released during the process of respiration (1).
(b) (the site of) protein manufacture (1)
(c) nucleus (1); which controls the functions of the cell (1); OR cytoplasm (1); where the cell reactions take place (1) OR cell membrane (1); controls movement of substances in and out of the cell (1).

3 The cell membrane is much thinner than the cellulose cell wall (1); the cell membrane controls the movement of substances in and out of the cell (1), but the cell wall is fully permeable (1); the cell wall has a role in supporting/strengthening the plant cell, but the cell membrane does not (because it is thin/flexible) (1).

2. Different kinds of cells

1 **(a)** cell wall/cell membrane (1). *Cytoplasm is a correct answer, but it is not shown in the diagram.*
(b) Bacterial cells differ from other cells in where their DNA/genetic material is found (1). In plant and animal cells this is found in a part of the cell called the nucleus (1). In bacterial cells this is found in plasmids/circular chromosomes in the cytoplasm (1).

2 **(a)** sperm cell (1)
(b) has a tail/has lots of mitochondria (1); so that it can move (1)
(c) It has no nucleus (1). *You may also have written that it is flexible/it has a thin cell membrane. This is correct, but not required by the specification.*
(d) more space for oxygen/haemoglobin (1) *Your answer here needs to match your answer to part (c). If you had one of the alternative answers in (c) then your answers here would have to be: it can squeeze into small blood vessels/oxygen can diffuse in and out very quickly.*

3 Any four from: yeast is a single-celled organism (1); the cells have a cell membrane (1); with an outer cell wall (1); inside the cytoplasm (1); there is a nucleus (1).

3. Diffusion

1 **(a)** Diffusion is the movement/spreading out of particles (1); so that there is a net movement from areas of high concentration (1) to areas of low concentration (1).
(b) They must be soluble (1).

2 **(a)** 8 μm +/−1 μm (1).
(b) oxygen (1)
(c) The first part of the statement is correct – diffusion will be fast when the concentration inside the cell is greater than the concentration outside (1); however, when the concentrations are the same diffusion does not stop (1); diffusion out of the cell and diffusion into the cell both happen, but at the same rate (1).

3 Particles of substance A diffuse in both directions, but more move to the right (1); from high to low concentration/until the numbers of particles is equal both sides (1); the number of particles of substance B is the same on both sides (1); substance B diffuses in both directions at the same rate/no net diffusion (1).

4. Organisation of the body

1 **(a)** organ (1)
(b) The lungs are composed of different types of tissues (1); each of these will be composed of a group of cells with similar structures and functions (1).
(c) oxygen or carbon dioxide (1)

2 **(a)** Differentiation is when cells become specialised (1); in order to carry out a specific function (1).
(b) They are long (1); so that the impulses can be carried long distances (1) OR it has branches at one end (1); to help communicate with other cells (1) OR it has an insulating layer (1); to help it conduct the impulse better (1).

3 A tissue is a collection of the cells of the same type (1); and that have a similar function (1). Any example from the following: muscle tissue (1), this contracts to enable organisms to move (1) OR glandular tissue (1), this produces enzymes/hormones (1) OR epithelial tissue (1), this covers parts of the body for protection (1).

5. Organs and organ systems

1 **(a)** organ (1)
(b) The heart is mostly composed of muscle tissue (1); this tissue can contract/beat to help move the blood around the body (1).
(c) to allow transport of blood/allow exchange of oxygen in respiring cells (1)

2 **(a)** a group of different tissues (1); that work together to perform a function (1)
(b) Any four of the following: muscular tissue (1), to help churn/mix the stomach contents (1); glandular tissue (1), to produce enzymes/protease/digestive juices (1); epithelial tissue (1), to cover the lining of the stomach (1).

3 Any four from: digestion of food takes place in the stomach and small intestine (1); enzymes/digestive juices break down the food into smaller molecules (1); these small food molecules are absorbed in the small intestine by diffusion (1); in the large intestine, water is absorbed from undigested food (1); the remaining faeces are then returned to the environment (1).

6. Plant organs

1 **(a)** The leaves of a plant are adapted to carry out photosynthesis (1); in order to provide glucose/sugars for the plant (1).
(b) Roots absorb water (1); roots absorb minerals and nutrients from the soil (1). *You could also say that roots help to anchor the plant in the ground, but this is not expected knowledge.*

2 A plant leaf is an organ because contains different types of tissue (1); these tissues include epidermal/mesophyll/xylem/phloem (1). *Your answer should give at least two of these types of tissue.* In an organ, the tissues work together to allow the leaf to perform the function (1); and in a leaf, this function is photosynthesis (1).

3 (a) It covers/protects the cells in the stem (1).
(b) xylem (1), phloem (1); transports substances around the plant (1)
(c) support for the leaves/holds leaves towards the sunlight/keeps leaves off the ground to prevent them being eaten by worms or slugs (1)

7. Photosynthesis

1 (a) oxygen (1)
(b) The greater the distance between the lamp and the plant, the lower the light intensity (1); as light intensity goes down, the rate of photosynthesis goes down (1).
2 (a) starch (1)
(b) cellulose (1); used to making the cell wall (1) OR protein (1); used for making new cells (1)
3 (a) mesophyll cells (1); contain chloroplasts (1); which are used to trap the light energy from the Sun (1)
(b) **dissolved** carbon dioxide (1); diffuses into the leaves of the pondweed (1)

8. Limiting factors

1 (a) 2.0 (1). *But anything between 1.9 and 2.1 will gain credit.*
(b) As temperature rises, the rate of photosynthesis increases (1). However, when the temperature rises further, the increase in the rate slows down/the rate does not increase as much (1).
2 The gradient of the graph is decreasing, indicating that the rate is approaching a maximum level (1); the rate of photosynthesis is being limited (1); by another factor, such as carbon dioxide concentration (1).
3 (a) It increases it (1).
(b) Increase the concentration of carbon dioxide in the greenhouse (1); the levelling off of the graphs shows that the rate is being limited (1); and this cannot be by temperature or light intensity, so must be due to carbon dioxide (1).

9. Biology six mark question 1

A basic answer will contain a general comparison of the cells and may lack detail. Only a few cell structures are identified and little detail on the function of each structure is given.
A good answer will compare at least two of the cells in detail. A fairly complete list of cell structures will be identified, some with functions.
An excellent answer will contain a complete, detailed and balanced comparison of all three cells. There should be a complete list of cell structures, correctly used to identify the cell types. These structures will also be given a function.

Examples of biology points made in the response:

- A = animal cell, B = bacteria, C = plant cell.
- All cells have a cytoplasm and a cell membrane.
- Cytoplasm acts as the medium for cell reactions.
- Cell membrane controls substances coming in and out of the cell.
- Bacteria cell has a cell wall, but no distinct nucleus.
- Animal and plant cells have a nucleus, mitochondria and ribosomes.
- Nucleus controls cell activity.
- Mitochondria are where energy is transferred in respiration.
- Ribosomes are where proteins are made.
- Additionally, plant cell has chloroplasts and vacuole.
- Chloroplasts are site of photosynthesis.
- Vacuole is filled with sap and helps keep cell shape.

10. Distribution of organisms

1 (a) 8 in dark side (1); 2 in light side (1)
(b) The student would need to put the same number of woodlice on each side at the start of the experiment (1); this makes sure that the test is not biased/the experiment is not affected because all the woodlice have been put on the same side (1).
(c) The student's experiment shows that woodlice prefer living in dark conditions (1); and in damp conditions (1); these conditions are going to be found in the habitat they live in, under rotting tree trunks (1).
2 More lichens grow on the tree than on the concrete post (1); as the tree can provide nutrients to the lichen/the concrete post has no nutrients to pass on to the lichen (1); the greater the light intensity, the greater the growth of lichen (1); showing that lichens also need light in order to grow (1).

11. Sampling organisms

1 (a) He could use a 1 m × 1 m quadrat (1); which he would throw at different places around the flower bed (1).
(b) Using the same area means that his experiment is valid/a fair test (1).
(c) He could look at more than one area each day (1); and take an average number of slugs (1).
2 (a) Use a quadrat several times in different parts of the field (1); and calculate a mean number of clover plants (1).
(b) total size of field = 100 × 65 = 6500 m^2 (1); so number of clover plants = 6500 × 7 (1) = 45 500 (1)
3 The mode is the number that appears most frequently (1); in this case, it's 12 (1); the mean is found by adding the data and dividing by the number of pieces of data used (1); which is (11 + 12 + 7 + 12 + 8 + 8 + 12)/7 = 10 (1); the median is the middle number when arranged in numerical order (1); which is 11 (1).

12. Transect sampling

1 (a) A transect can be used to sample the distribution of organisms (1); between two neighbouring habitats or areas (1); here, the scientist could draw a line from the sea shore up the beach, perpendicular to the sea line (1).
(b) (10 + 8 + 9)/3 (1) = 9 (1)
(c) the number of limpets goes down as you travel further from the sea (1); this decrease is linear with the distance/it drops by 4 limpets for every 0.5 m distance travelled (1); limpets therefore prefer to live nearer the seashore (1).
2 (a) The scientist should do another transect line into the wood and place quadrats at the same distance (1); so that she can calculate a mean value (1).
(b) Any four from: the number of bluebells increases to a maximum around 8 m into the wood (1); and then decreases as you go deeper into the wood (1); this is because light intensity decreases as you deeper into the wood (1); so less light available for photosynthesis (1); and fewer nutrients/water available deeper in the wood where there are more trees (1).

13. Proteins

1 (a) enzymes (1)
(b) If it gets too warm, the protein molecule can denature/unfold (1). This changes the shape of the molecule, making it less effective (1).

2 (a) from the diet/from food (1)
(b) Antibodies are proteins (1); and they destroy pathogens (1).
(c) Protein needed to help build muscles (1); so weightlifters can build muscle tissue more quickly if they take the supplement (1).

3 A series of amino acids (1); bind together to form a long chain (1); this chain folds (1); to produce a specific shape/ to allow the iron atom to fit exactly inside (1).

14. Digestive enzymes

1 (a) proteases (1)
(b) They catalyse/speed up (1); the breakdown of protein molecules (1); into amino acids (1).

2 (a) The pancreas produces a variety of digestive enzymes (1); these are released into a part of the digestive system called the small intestine (1).
(b) fatty acids (1); and glycerol (1)
(c) Churned food leaving the stomach is acidic (1); enzymes in the small intestine only work in alkaline conditions (1); so the acid needs to be neutralised (1); body produces bile to help achieve this (1).

3 Up to around 37 °C, activity increases because the temperature is increasing (1); activity reaches a maximum around 37 °C as this is the optimum temperature (1); above this temperature the amylase denatures/changes shape/ becomes inactive (1); and so the activity decreases (1).

15. Microbial enzymes

1 (a) lipase (1)
(b) Blood contains proteins (1).
(c) Some stains may contain more than one type of substance/clothing may have more than one stain (1); so a single enzyme would not remove all the stains (1).

2 (a) The peel must contain protein (1); because the protease enzymes break it down (1).
(b) The enzymes would soak further into the oranges (1); and break up proteins that make up the 'flesh' of the fruit (1).

3 Advantages: proteases are used to help pre-digest protein in baby foods (1); isomerase is used to convert glucose into fructose, which is sweeter (1); this is very useful in making foods for slimmers (1); using enzymes allows reactions to take place at much lower temperatures (1) thus avoiding the use of expensive equipment (1). Disadvantages: many enzymes are costly to produce (1); enzymes can be easily denatured by changes in temperature/pH (1); there might be health concerns/allergic reactions (1). Conclusion: this will balance the advantages and disadvantages, for example: 'despite the disadvantages, the potential use of enzymes in the food industry outweighs them' (1). *There are five marks available for advantages and disadvantages of the use of enzymes. Your answer must include at least one advantage and at least one disadvantage. The final mark is for your conclusion.*

16. Aerobic respiration

1 (a) glucose + oxygen (1) → water + carbon dioxide (+ energy) (1)
(b) releasing energy from food/sugars (1)
(c) Respiration is slower at night than in the day (1) (*Note that cells will respire at night – so if you said 'Cells only respire in the day' this is not correct*); because body is more active during the day/body is resting at night (1).

2 (a) To make muscles contract/allowing animals to move (1); maintain a steady body temperature (1).
(b) Plants use energy to make sugars (1); these are combined with nitrates/nutrients to make amino acids (1); which then make proteins for growth (1).

3 (a) Muscle cells need to expend more energy/help us to move (1); so need more energy from respiration, which happens in the mitochondria (1) OR Reverse argument for skin cells: skin cells do not move, so do not need to expend energy (1); so need few mitochondria, as this is where release of energy through respiration occurs (1).
(b) Respiration uses enzymes (1); and these enzymes are easily denatured by higher temperatures (1).

17. The effect of exercise

1 (a) The graph shows that exercise increases the breathing rate (1). However, after a few minutes, the breathing rate levels out/stops increasing (1).
(b) 35 − 15 (1) = 20 (1) breaths per minute (1)
(c) in 60 seconds, number of breaths = 60/4 (1) = 15 breaths per minute (1)

2 (a) Heart rate would increase (1); the exercise uses the muscles, so more blood will flow to them (1).
(b) Glucose carried in the blood to the muscles (1); enters the muscle cells by diffusion (1); some glucose is stored as glycogen in the muscle cells (1); this is converted back into glucose in the cells (1).

18. Anaerobic respiration

1 (a) They become fatigued (1).
(b) A chemical called lactic acid (1); starts to build up/ increase in concentration in her muscles (1).

2 Any four from: both processes use up glucose (1), but aerobic respiration also uses oxygen (1); anaerobic respiration produces lactic acid (1), but aerobic respiration produces carbon dioxide and water (1); anaerobic respiration happens when oxygen cannot be supplied to tissues quickly enough (1).

3 Any six from: oxygen consumption increases during exercise (1); reaches a maximum value when no more oxygen can be supplied to cells (1); anaerobic respiration begins to take place (1); producing lactic acid (1); after exercise stops, oxygen consumption remains high (1); this oxygen is needed to convert lactic acid into carbon dioxide and water (1); often referred to as an 'oxygen debt' (1).

19. Biology six mark question 2

A basic answer would mostly make observations based on the graph without relating the changes to what is happening in the body.
A good answer would give a link between the increase in heart and breathing rate and the need for more blood flow and more oxygen to respiring cells. Some use is made of data from the graphs.
An excellent answer would give, in addition, information about the different types of respiration, and what is happening in the recovery phase and why. There is good use of data from the graphs, such as the change in breathing rate identified between sections.

Examples of biology points made in the response:

- Graph shows constant breathing rate and heart rate in initial rest period.

- Breathing rate and heart rate increase with exercise.
- This is shown by lines on the graph rising.
- This increase is due to increased blood flow to muscles; this increases delivery of oxygen and sugar/glucose and increases removal of carbon dioxide.
- Aerobic respiration occurs if cells get enough oxygen.
- Change from rest to light exercise is smaller than change from light exercise to heavy exercise.
- Change in heart rate from rest to light exercise is about 25 bpm.
- Heart rate rises another 50–75 bpm in heavy exercise.
- Graphs start to flatten off as this maximum rate is reached.
- No extra oxygen is therefore getting supplied to cells.
- Likely that cells will not have all the oxygen they need, so may respire anaerobically as well as aerobically.
- This produces lactic acid.
- Muscles become fatigued after period of exercise.
- Even when exercise stops, they need increased blood flow to remove metabolic products.
- Therefore, the lines on the graph do not return immediately to the resting level before exercise.
- **Need increased oxygen flow to help oxidise lactic acid.**

The last point is part of the Higher Tier specification so would not be expected if you are entered for the Foundation Tier.

20. Mitosis

1 At 2 hours: the 1 parent cell divides by mitosis to give 2 daughter cells. At 4 hours: the 2 cells each divide by mitosis to give 4 new cells (1). At 6 hours: the 4 cells each divide by mitosis to give 8 new cells (1).

2 **(a)** The daughter cells are genetically identical to the parent cell (1).
(b) two daughter cells (1); both cells with the same number of chromosomes as the parent cell (1)

3 **(a)** producing new cells for growth (1); repairing damaged cells (1)
(b) Any four from: graph rises after 12 hours as cell starts making DNA (1); the cell is starting to enter mitosis by copying genetic material (1); graph levels off when amount of DNA has been doubled (1); just before 24 hours, the amount of DNA in the cell halves (1); because the cell has separated into two daughter cells (1).

21. Meiosis

1 **(a)** Each human body cell contains a total of 46 chromosomes (1). These chromosomes are normally found in pairs in the cells (1).
(b) One pair of chromosomes are the X and Y chromosomes (1); if a child inherits an X chromosome from both parents/is XX then they are a girl (1); if they inherit X from mother and Y from father/is XY then they are a boy (1).

2 four daughter cells (1); two cells with one long chromosome and the other two cells with one short chromosome (1)

3 **(a)** Sperm cell and egg cell fuse (come together) (1); in a process called fertilisation (1).
(b) To make sex cells, the process is meiosis (1); this produces gametes (1), which have only one set of chromosomes/half the number of chromosome as body cells (1); as the foetus develops, new cells are produced for growth through mitosis (1); and the cells produced have 46 chromosomes/two sets of chromosomes (1).

22. Stem cells

1 **(a)** In differentiation, stem cells become specialised (1); to carry out a specific function (1).
(b) Plant stem cells can differentiate throughout their life cycle (1); whereas animal stem cells differentiate only early in their life cycle (1); both types of stem cell can differentiate into cells of many different types (1).

2 **(a)** $1250 - (775 + 367)$ (1) $= 108$ (1)
(b) $775/1250 \times 100\%$ (1) $= 62\%$ (1)
(c) As scientists begin to develop new treatments based on embryonic stem cells (1); people may begin to see the benefits of the research outweigh the negatives (1).

3 Any five from: adult stem cell treatment has good success rate/40% success rate (1); stem cells from the blood can differentiate into many different types of cell (1), so the possibilities for treatments are large (1); embryonic stem cells offer the same potential for treatments (1); although there are ethical concerns about using embryos for this purpose (1); and there are greater medical differences as patients may have a reaction against the foreign embryo tissue (1); there may be social concerns about any stem cell research as it implies that people with paralysis are 'unhealthy' (1); the size of the sample (30 patients) is very small, so may not give reproducible data (1). For the sixth mark you must give an evaluation/conclusion, which considers both sides, for example: 'Although some people disagree with stem cell treatments, the potential they have is very great'; or 'Although there are potential cures available through this research, it is at the expense of destroying precious embryos'; or 'There are fewer ethical concerns about using adult stem cells, and it is effective, so this is better than using embryonic stem cells.'

23. Genes and alleles

1 **(a)** DNA consists of two strands of deoxyribonucleic acid (1); arranged together in a double helix (1).
(b) A gene is a short section of DNA (1); a chromosome is a long DNA molecule, containing many genes (1).

2 **(a)** Alleles are different forms of a gene (1).
(b) height/hair colour/shape of ears (1)

3 **(a)** A gene's code is used to make combinations of amino acids (1); which then produce a specific protein (1).
(b) Any four from: the twins have the same genetic composition (1) *(Remember that you are not expected to know a reason for this)*; the twins have a different genetic composition to each of their parents (1); because they have inherited a combination of alleles from each parent (1); because the twins are produced through sexual reproduction (1); and this process leads to variation (1).

24. Genetic diagrams

1 **(a)** **dd** (1); the allele is recessive, so two copies needed in order to have badly formed wings (1).
(b) **dd** have badly formed wings and both **DD** and **Dd** have normal wings (1); so ratio of normal to badly formed wings is 3-to-1/0.75/75% (1).

2 **(a)** the allele **B** is dominant(1); so mice with the genotypes **BB** (1) or **Bb** (1) have black fur.
(b) **(i)** Heterozygous means the two alleles are different from each other, but homozygous means that they are the same (1).

(ii) In the diagram, the initial genotypes should be **Bb** and **bb** (1); giving rise to the alleles as shown in the diagram (1); and the genotypes of the offspring (1); giving a ratio of 50% black : 50% brown in phenotype (1).

		Father **Bb**	
		B	**b**
Mother **bb**	**b**	**Bb**	**bb**
	b	**Bb**	**bb**

25. Mendel's work

1 (a) This allowed Mendel to collect several sets of data (1). This meant that his results were likely to be reproducible (1).

(b) This meant that Mendel knew exactly which the parent plants were (1); so that he could show how the characteristics were inherited from parent to offspring plant (1).

(c) These plants always behave the same way when crossed together (1); so that there was no random variation (1).

2 Any pair of reasons from the following: it is not ethical to do this work with humans (1), because we have free will/are more complex (1). You could not reproduce this work with humans (1), as we are genetically more complex (1). The reproductive cycle for humans is too slow (1) to be able to do all the work in the time (1). This sort of inter-breeding with humans (1) could lead to genetic problems (1).

3 (a) **RR** – red flowers (1); **Rr** – red flowers (1); **rr** – white flowers (1)

(b) use of genetic diagram/Punnett square to show that the genotypes of the offspring are **RR**, **Rr**, **Rr** and **rr** (1); so ratio should be 3 : 1 (1); which is approximately the same as the ratio seen of 76 : 24 (1)

26. Punnett squares

1 (a) **P** (1)

(b) The probability of having a child with polydactyly is 50% (1); because children with the alleles **Pp** show polydactyly but those with **pp** do not (1).

2 (a) Only one copy is need in order to develop the condition (1).

(b) **aa** (1)

(c) All of the gametes he produces will have the **A** allele (1); so offspring all inherit this allele from their father (1); as it is a dominant allele, they have the condition even if they inherit **a** from the mother (1).

3 (a) initial genotypes of the parents as **PP** and **pp** (1); the genotypes of all offspring as **Pp** (1)

		Parent **PP**	
		P	**P**
Parent **pp**	**p**	**Pp**	**Pp**
	p	**Pp**	**Pp**

(b) Because the **P** allele is dominant (1); people that are heterozygous will have the condition, rather than just acting as carriers (1).

27. Family trees

1 (a) The condition is only seen if an individual inherits two copies of the allele for cystic fibrosis (1).

(b) The man has the alleles **Ff** (1). He is able to pass on the allele for cystic fibrosis (1); although he does not have the condition himself (1).

2 (a) 50% (1); those who are **bb** will have Batten disease/those who are **Bb** do not have Batten disease (1).

(b) Mother now passes on **B** allele only/children cannot inherit two copies of **b** allele (1), so they cannot develop Batten disease (1).

3 (a) Children with **hh** will show sickle cell anaemia but those with **HH** or **Hh** will not (1); so only 1 in 4 of the offspring (25%) will show the condition (1).

(b) Both parents are carriers for the condition (1); so they both have the allele for sickle cell anaemia (1); if they both pass on this allele to their child, then the child can develop the condition (1).

28. Embryo screening

1 (a) 2 (1)

(b) (i) The test shows that the embryo has Down syndrome, when it does not (1)

(ii) parents face difficult decisions, so need the test to be correct (1); so that they can be confident that the decision they make is based on what is really happening (1).

(c) The graph shows that the chance of having a child with Down syndrome increases as the mother's age increases (1); therefore pregnant mothers over the age of 40 – who are at greater risk – are strongly advised to have the embryo screened (1).

2 (a) Up to three marks for points made about embryo screening: screening is useful because: it allows you to prepare for having a child with HFI (1); it allows you to make a decision about terminating the pregnancy (1); there may be economic factors in having a child with HFI because of diet or medication (1). Screening is not useful for HFI because: the condition is not life-threatening (1); parents are unlikely to consider terminating a pregnancy based on this condition alone (1). A fourth mark is for an evaluation as part of the conclusion, for example: 'Although embryo screening is important, this condition is easy to manage so not important to screen for it'; or 'Although the condition is not life-threatening, it reduces quality of life so it is important to screen for it.'

(b) Slimming foods contain a great deal of fructose (1); as it is sweeter than glucose (1).

29. Fossils

1 (a) As the scientists dig down further, the fossils they find are older (1).

(b) The conditions do not allow the organism to decay (1). *You could also have said low oxygen levels/high acidity.*

(c) Bone/teeth/hard parts of the organism (1); do not easily decay/decay less easily than soft tissues like skin (1).

2 (a) More recent fossils have a smaller jaw (1); larger/longer skull (1).

(b) human (1); from shape of skull/face/pattern of teeth (1)

3 The fossil record is not always complete (1). Then any two from: because the conditions may not be right for fossils to form (1); or the remains have been destroyed by geological changes in the Earth (1); or the organism simply decays without fossilisation (1); or it is rare to find that any fossil record goes back as far as the start of life on Earth (1); *plus* so there is not enough valid/reliable evidence for scientists to make any conclusions about the start of life on Earth (1).

30. Extinction

1 **(a)** Christmas Island rats were isolated/the island is not connected to the mainland (1); so the rats there evolved in a different way (1); to form a new species (1).
(b) **(i)** The rats were extinct in 1908, after the introduction of black rats in 1899 – this is 9 years (1).
(ii) The black rats carried a disease (1); which they were immune to, but the native rats were not (1). *There are other reasons this could have happened – such as the black rats outcompeting the native rats – but the question mentioned a parasite, so this is the expected answer.*

2 **(a)** **(i)** Many species die out at the same time (1).
(ii) Any three from: volcanic eruption/collision of an asteroid with Earth (1); releasing dust into the atmosphere (1); blocks out sunlight/causes a sudden change in climate (1); plants can no longer photosynthesise/plants and animals lose food sources (1).
(b) Mammoths and humans existed at the same time/ shared the same habitats (1); so we do not have any evidence for a catastrophic event killing off many species (1).
(c) Extinction was through hunting (1); as humans needed their fur/ivory/meat (1).

31. New species

1 **(a)** Land divides porkfish population into two (1); so fish on each side of the land can no longer mate together (1).
(b) Speciation is the development of new species (1); that cannot interbreed with each other (1).

2 Any four from: variation exists between different individuals in a species (1); finches were separated and ended up on different islands (1); those individuals best suited to the environment on each island survived (1); these organisms bred more successfully (1); and passed on their characteristics so that each type of finch developed differently (1).

3 **(a)** speciation (1)
(b) Any four from: there are physical differences between salamanders in a particular area (1) because their alleles are expressed in a different way/they have different combinations of alleles (1); this can lead to natural selection (1); as a particular characteristic may give some salamanders an advantage in their habitat (1); the salamanders in different areas therefore become more different from each other (1).

32. Biology six mark question 3

A basic answer would give a collection of facts about the different processes taking place but none of the three processes will be described in much detail.
A good answer would give a fairly good description of one or two of the different cell processes taking place; the weak answer is likely to be on differentiation to form the fetus.
An excellent answer would give a good description of all three processes, with clear, correct use of the technical words provided in the guidance box.

Examples of biology points made in the response:

- Sperm and egg gametes made through meiosis.
- They fuse together in fertilisation.
- Single body cell (zygote) formed with new chromosomes from the gametes.
- Zygote divides by mitosis, forming identical cells.
- In mitosis, DNA is replicated to make two copies and each daughter cell takes one copy of the DNA.
- Number of cells doubles with each mitotic division.
- Daughter cells are genetically identical.
- Each cell in the early embryo is a stem cell.
- Stem cells can differentiate to become all specialised cells of the body.
- Embryo continues to divide by mitosis and cells differentiate to form organs in the fetus.

Chemistry answers

33. Forming ions

1 Missing data: Ca, Ca^{2+}, O, O^{2-} (2)

2 **(a)** B (1)
(b) A (1)

3 **(a)** Element: a substance that contains only one kind of atom. Compound: contains two or more kinds of atom (1) joined together (1). Bond: a force of attraction that holds atoms together (1).
(b) any number between 90 and 120 (1)
(c) The two main types are covalent bonds (1) formed by sharing electrons (1) and ionic bonds (1) formed by transferring electrons (1).

4 **(a)** Mg^{2+} (1)
(b) Magnesium loses the 2 electrons (1) in the outer shell (1).
(c) beryllium or calcium (1)

34. Ionic compounds

1 **(a)** a regular arrangement (1) containing billions of atoms/ions (1)
(b) metals and non-metals (1)
(c) Sodium atoms each lose an electron to form positive ions (1) and chlorine atoms each gain an electron to form negative ions (1). The positive and negative ions attract each other (1) to form the lattice structure.

2 **(a)** $ZnBr_2$ (1)
(b) ZnS (1)
(c) K_3N (1)

3 **(a)** alkali metals (1)
(b) $2Na + F_2$ (1) $\rightarrow$ 2NaF (1)
(c) 3+ (1)

4 **(a)** halogens (1)
(b) Both fluorine and chlorine have 7 electrons in their outer shell (1). Therefore both need to gain an outer electron to become stable (1). To form an ionic bond one atom needs to lose electrons while the other atom gains electrons (1).

35. Giant ionic structures

1 **(a)**

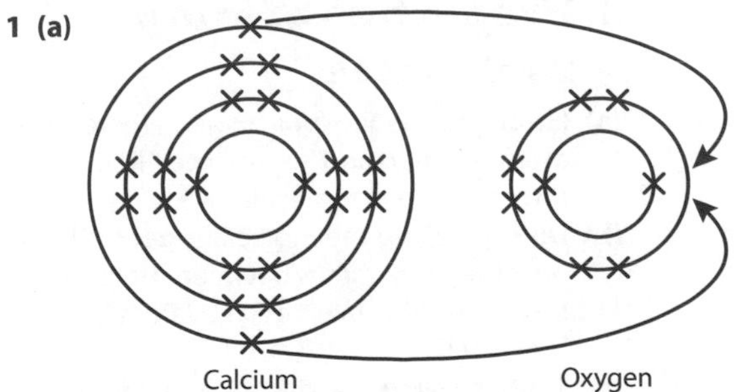

Arrows showing two outer electrons from calcium moving to the outer shell of the oxygen atom (1).

(b) The calcium atoms lose 2 electrons (1) which gives them a full outer shell like a noble gas/argon (1).

(c) Ca^{2+} (1); O^{2-} (1)

2 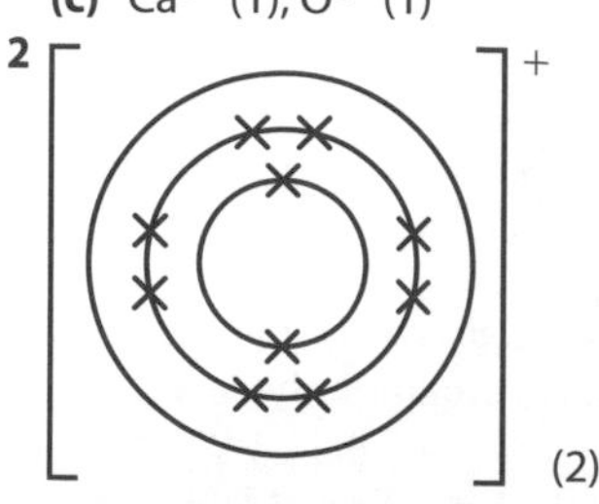

(2)

3 Magnesium atoms lose 2 electrons (1) and chlorine atoms gain 1 electron (1). Therefore 2 atoms of chlorine are needed for each magnesium atom so an equal number of electrons are lost and gained (1).

36. Covalent bonds in simple molecules

1 (a) non-metal elements (1)

(b) Molecules are small groups of atoms (1) held together by covalent bonds. Giant covalent structures contain billions of atoms in a regular structure/lattice (1).

2 (a) carbon hydride/methane (1); CH_4 (1)

(b) (i) covalent bonds (1)

(ii) formed by sharing electrons (1)

(c) A molecular structure has groups of atoms (1) held together by covalent bonds (1).

3 hydrogen oxide/water H_2O (or carbon dioxide CO_2) (1) and silicon oxide/silicon dioxide SiO_2 (1)

37. Covalent bonds in macromolecules

1 (a) (i) The lines represent shared pairs of electrons (1), which form the covalent bonds (1).

(ii)

```
    H
    |
H — Si — H
    |
    H
```

(1)

(b) Diagram like the one to the right, showing two electrons shared between the atoms (1) and showing the chlorine atom with 6 further electrons (1).

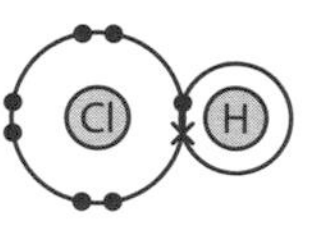

2 (a) N_2 (1)

(b) 3 (1)

3 Diagram like the one to the right, showing 3 pairs of electrons shared between the nitrogen atom and three hydrogen atoms (1), and showing the nitrogen atom with 2 further electrons (1).

```
  H
  •×
: N •× H
  ×•
  H
```

38. Properties of simple molecules

1 (a) 'Strong forces' line pointing to where circles of the atoms overlap and 'weak forces' line pointing to forces between molecules (1).

(b) During boiling the molecules get further apart (1); only the weak forces between molecules need to be broken (1). The strong covalent bonds are not broken (1).

(c) Freely moving charges are needed to conduct electricity (1), and covalent substances have no freely moving charged particles (1).

2 (a) Diagram like the one below, showing carbon rod (1), light bulb (1), power supply (1), and switch (1).

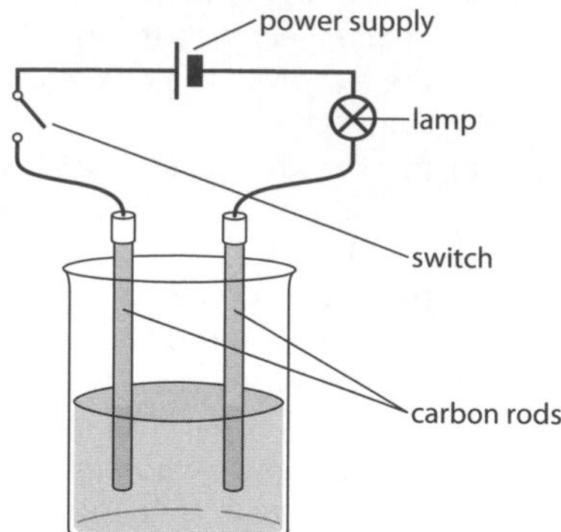

(b) Solutions A, C and D contain ions, or A, C and D are ionic (1). B is covalent/molecular, or the solution of B contains no ions (1).

39. Properties of macromolecules

1

Small molecules (1)	Giant covalent (or covalent lattice) (1)
carbon dioxide	diamond
water (1)	silicon dioxide (1)
hydrogen chloride (1)	graphite (1)

2 (a) Strong covalent bonds (3 per atom) hold atoms in layer structure (1). There are also weak bonds/intermolecular forces between the layers (1).

(b) The layers can slip over each other (1) as the forces between the layers are weak (1).

(c) Only 3 bonds per atom in graphite (1), leaves 1 outer electron on each carbon atom free to move (1) (and conduct electricity), but in diamond there are no free electrons or all electrons are used in covalent bonds (1).

(d) In graphite and fullerene all carbon atoms form 3 bonds (1). Graphite is made up of sheets of atoms (containing many atoms) (1). Fullerene forms molecules (of set size) (1).

3 (a) covalent (1)

(b) Carbon dioxide is a gas at room temperature as it is made up of molecules (1) and only weak forces between the molecules have to be broken (1) to boil it. Silicon dioxide is a high-melting-point solid as it has a lattice structure/giant covalent structure (1) and billions of strong covalent bonds have to be broken (1) to boil it.

40. Properties of ionic compounds

1 (a) ionic (1)

(b) electrostatic forces (1) due to attraction between positively and negatively charged ions (1)

(c) Solid X does not conduct electricity because the ions cannot move (1). A solution of X conducts electricity as the lattice is broken up (1) and the ions can move freely (1).

2 (a) $2Na + Cl_2$ (1) $\rightarrow$ $2NaCl$ (1)

(b) It has a lattice structure (1) and billions of strong bonds need to be broken to melt it (1).

3 (a) $BaCl_2$ (1)

(b) Barium atoms lose 2 electrons (1) and chlorine atoms gain 1 electron (1).

(c) Ionic compounds like barium chloride have a lattice structure (1) containing billions of oppositely charged ions (1), which are held together by the attractive force between the opposite charges (1).

4 Any four from: lithium chloride is made up of a regular lattice containing oppositely charged ions (1). It has a high melting point (1) as large amounts of energy are needed to break the many strong bonds (1). It conducts electricity when molten or dissolved in water (1) as the ions are free to move (1) OR it does not conduct electricity when solid (1) as the ions cannot move freely (1).

41. Metals

1 **(a)** A shape memory alloy can return to its original shape (1) after being bent/deformed (1).
(b) nitol (1), used in dental braces (1). *In an exam you would get credit for any other correct name and use, but the one given here is the only one you are* ***expected*** *to remember.*

2 **(a)** Their outer electrons (1) can move through the structure (1).
(b) As the temperature rises the conductivity falls steeply (1) at first and then starts to level off (1).
(c) Any two of: low density/light (1); strong (1); resists corrosion (1).

3 **(a)** a mixture of metals/a metal mixed with another element (1)
(b) lattice (1)
(c) The differently sized atoms distort the layers of metal atoms (1) making it more difficult for the layers to slide over each other (1).

42. Polymers

1 **(a)** Polymers consist of long chain molecules (1) made by joining small molecules together (1)
(b) down pipes: don't rust/lighter (1); food bags: stronger/waterproof (1)

2 **(a)** Thermosoftening polymers melt on heating (1), whereas thermosetting polymers do not melt on heating (1).
(b) Thermosetting polymers are better for pot handles (1) as they remain rigid (1) when they get hot.
(c) Thermosoftening polymers have long unattached molecules (1), which can move away from each other (1) and melt. Thermosetting polymers have links between the molecules (1), which means they cannot move about easily (1).

43. Nanoscience

1 **(a)** very small particles (1 to 100 nm) (1)
(b) The surface area of nanoparticles is very large compared with the particles in ordinary powders (1).

2 **(a)** because they are so small they can get to places other particles couldn't (1)
(b) large surface area (1)
(c) There are possible dangers of using nanoparticles (1) as we don't know all their properties (1).

3 **(a)** lithium (1); **(b)** poly(ethene) (1)

44. Chemistry six mark question 1

A brief answer would have a simple description of the bonding types in both sodium chloride and in hydrogen chloride.
A good answer would have a clear description of the bonding types and associated properties of melting point/boiling point, in both sodium chloride and in hydrogen chloride.
An excellent answer would have a detailed description of the bonding types in both sodium chloride and in hydrogen chloride, with an explanation of the associated properties of melting point/boiling point.

Examples of chemistry points made in the response:

- Sodium chloride contains ionic bonds (formed by transferring electrons to form ions).
- The ions are attracted to each other by electrostatic forces (oppositely charged ions attract) and form a giant regular structure (lattice structure).
- Ionic compounds are solids with high melting points, as large amounts of energy are needed to break the many strong bonds.
- Hydrogen chloride contains covalent bonding (formed by sharing electrons).
- It is made up of molecules.
- Substances that contain molecules have low melting points and low boiling points.

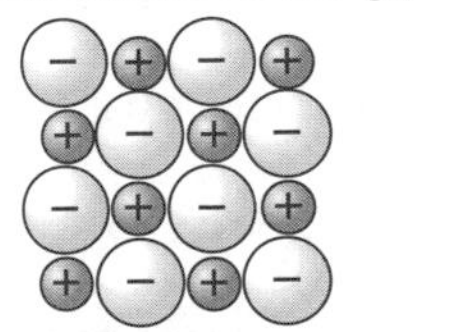

sodium chloride structure

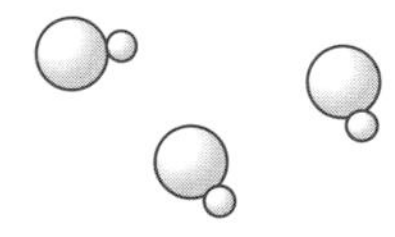
hydrogen chloride structure

45. Atomic structure and isotopes

1 **(a)** proton: positive, 1, in the nucleus (1); neutron: no charge, 1, in the nucleus (1); electron: negative, very small, around the nucleus (1)
(b) 21 protons (1); 21 electrons (1); 24 neutrons (1)

2 **(a)** one has 2 neutrons more than the other (1)
(b) $^{127}_{53}I$ (2)

3 Line 1: 28, 59, 28, 31, 28 (1). Line 2: $^{90}_{38}Sr$, 38, 90 (1). Line 3: $^{19}_{9}F$, 19, 9 (1).

4 **(a)** X = 82 protons, 125 neutrons, 82 electrons (1). Y = 83, protons, 124 neutrons, 83 electrons (1).
(b) They are not the same element (1) as they have different atomic numbers (1) (numbers of protons/electrons).

5 **(a)** isotopes (1)
(b) the average mass of the atoms of an element (1), relative to carbon-12 isotope (1)
(c) There is more $^{20}_{10}Ne$ than $^{22}_{10}Ne$(1).

46. Relative formula mass

1 **(a)** $(1 \times 16) + (2 \times 23)$ (1) $= 62$ (1)
(b) $(12 \times 12) + (22 \times 1) + (11 \times 16)$ (1) $= 342$ (1)
(c) $(4 \times 12) + (8 \times 1) + (2 \times 16)$ (1) $= 88$ (1)

2 **(a)** $(1 \times 31) + (3 \times 1)$ (1) $= 34$ g (1)
(b) $(2 \times 1) + (1 \times 32) + (4 \times 16) = 98$; 1 mole = 98 g (1); 2 moles = 2×98 g = 196 g (1)

3 **(a)** 1 mole $CO_2 = (1 \times 12) + (2 \times 16) = 44$ g (1); 88 g = 88/44 = 2 moles (1)
(b) 1 mole $O_2 = 2 \times 16 = 32$ g (1); 4 g = 4/32 = 0.125 moles (1)

47. Paper chromatography

1 **(a)** A: spot of purple food dye (1); B: paper (1); C: beaker/jar (1); D: solvent/water (1)
(b) The level of the solvent is below the paper (1), so the solvent cannot soak up the paper and separate the mixture (1).

2 **(a)** The dyes separated as each had a different solubility in the solvent (1) and so moved up the paper at different speeds/by different amounts (1).
(b) purple ink (1)

(c) It tells us the blue ink contains 2 dyes (1) and the black ink contains 3 dyes (1), and that 2 of the dyes are the same in each ink (1).
(d) The dye in the red ink (1), as it has moved furthest with solvent (1).

48. Gas chromatography

1 **(a)** The gas chromatography separates the mixture (1).
(b) The helium gas carries the substances in the mixture through (1).
(c) unreactive (1)
(d) In the coil the different substances in the mixture travel at different speeds (1) and separate from each other (1).
(e) measures the relative molecular mass of substances (1) and helps identify the substances (1)
2 **(a)** 6 (1)
(b) It travels quickest/it has the smallest relative molecular mass (1).
(c) their relative molecular mass (1)

49. Chemical calculations

1 **(a)** Mass of carbon $= 6 \times 12 = 72$;
% carbon $= 72/176 \times 100\%$ (1) $= 40.9\%$ (1)
(b) Mass of hydrogen $= 8 \times 1 = 8$;
% hydrogen $= 8/176 \times 100\%$ (1) $= 4.5\%$ (1)
2 **(a)** $39 + 14 + (3 \times 16) = 101$ (1);
% nitrogen $= 14/101 \times 100\% = 13.9\%$ (1)
(b) $14 + (4 \times 1) + 35.5 = 53.5$ (1);
% nitrogen $= 14/53.5 \times 100 = 26.2\%$ (1)
3 **(a)**

element	1 mole in g	mass in g	number of moles (1)	ratio (1)
C	12	4.5	4.5/12 = 0.375	0.375/0.375 = 1
H	1	1.5	1.5/1 = 1.5	1.5/0.375 = 4

Empirical formula CH_4 (1)

(b)

element	1 mole in g	mass in g	number of moles (1)	ratio (1)
S	32	3.2	3.2/32 = 0.1	0.1/0.1 = 1
O	16	8 − 3.2 = 4.8	4.8/16 = 0.3	0.3/0.1 = 3

Empirical formula SO_3 (1)

4

element	1 mole in g	%	number of moles (1)	ratio (1)
C	12	40	40/12 = 3.33	3.33/3.33 = 1
H	1	6.7	6.7/1 = 6.7	6.7/3.33 = 2
O	16	53.3	53.3/16 = 3.33	3.33/3.33 = 1

Empirical formula CH_2O (1)

50. Reacting mass calculations

1 **(a)** 1 mole of $CH_4 \rightarrow$ 1 mole CO_2 (1)
16 g → 44 g (1); 1 g → 44/16 g = 2.75 g CO_2 (1)
(b) 1 mole of CaO gives 1 mole $Ca(OH)_2$ (1);
56 g gives 74 g (1)
1 g 74/56 = 1.32 g; 1 kg gives 1.32 kg
5000 → 5000 × 1.32 = 6607.14 kg $Ca(OH)_2$ (1)
2 2 mole of $Fe_2O_3 \rightarrow$ 4 moles Fe (1)
2 × 160 g = 320 g → 4 × 56 g = 224 g (1)
1 g → 224/320 = 0.7
2000 → 2000 × 0.7 = 1400 g Fe (1)
3 1 mole of Na_2SO_4 needs 2 moles NaOH (1)
142 g needs 2 × 40 g = 80 g (1) 1 g needs 80/142 = 0.563
25 needs 25 × 0.563 = 14.08 g NaOH (1)

51. Reaction yields

1 **(a)** During a reaction the atoms of the reactants are rearranged to form the products (1); no atoms are lost or gained (1).
(b) Reaction might not go to completion (1). Some product might be lost (1). Some reactants may react in a different way (1).
(c) $\%\text{ yield} = \frac{\text{actual yield}}{\text{calculated yield}} \times 100\%$ (1)
2 **(a)** 0.8 g (1); **(b)** 1 g (1)
(c) Some of magnesium oxide is lost (1) as there is no lid (1). Some of the magnesium might not react completely (1) as there is not a good supply of oxygen (1).

52. Reversible reactions

1 **(a)** This means the reaction goes in both directions (1).
(b) carbon monoxide, hydrogen (1), methane and water (1)
2 16 × 20/100 (1) = 3.2 g of methane (1)
3 **(a)** Nitrogen and hydrogen are recycled (1), and this makes the process sustainable as no reactant is wasted (1).
(b) **(i)** 1 mole of $N_2 \rightarrow$ 2 mole NH_3 (1)
∴ 28 g → 2 × 17 g = 34 g (1)
∴ 1 g → 34/28 = 12.14
∴ 72 → 72 × 12.14 = 87.43 g NH_3 (1)
(ii) 87.43 × 5/100 (1) = 4.37 g (1)

53. Rates of reaction 1

1 **(a)** **(i)** rate = change/time = (0.36 − 0.22)/(4 − 2) (1)
= 0.14/2 = 0.07 g/min (1)
(ii) rate = change/time = (0.42 − 0.36)/(6 − 4) (1)
= 0.06/2 = 0.03 g/min (1)
(b) The reaction is completed in 7 minutes (1) as no more gas is produced/no more mass is lost after this point (1).
2 **(a)** rate = change/time = (30 − 0)/(20 − 0) (1) = 30/20 = 1.5 cm^3/s (1)
(b) rate = change/time = (15 − 0)/(20 − 0) (1) = 15/20 = 0.75 cm^3/s (1)
(c) 70 seconds (1)
(d) They don't use the same amount of reactants (1) as both reactions are completed in the time shown by the graph (1) and experiment A produces more carbon dioxide (1) (more reactants must have been used in experiment A).

54. Rates of reaction 2

1 **(a)** The reacting molecules have to collide (1) with enough energy (1).
(b) The rate decreases (1) as there are fewer molecules in the same volume of gas (1) and so fewer collisions will occur (1).
2 **(a)** **(i)** Flask B (1)
(ii) Two of: it has the lowest concentration of acid (1), lowest temperature (1) largest particle size (1) and least reactive metal (1).
(b) **(i)** E and F (1)
(ii) volume of acid (1) and mass/moles of metal (1)

55. Chemistry six mark question 2

A brief answer would have a simple description of yields and why some reactions do not produce a 100% yield.
A good answer would have a clear description of yields and the problems of low yields in industry. It should also include one or two reasons why some reactions do not produce a 100% yield.

An excellent answer would have a detailed description of yields and the problems of low yields in industry. It should also include two reasons why some reactions do not produce a 100% yield and a description of at least one way that the problem of a low yield can be overcome.

Examples of chemistry points made in the response:

- A low yield means that less product is formed than expected (by calculation).
- This is a problem in industrial processes as expensive reactants/feedstock/raw materials can be wasted (this can make the product very expensive as reactants are wasted).
- The main reasons for reactions producing less than 100% yields are that the reaction is reversible and products may be lost in separation (also some reactants may react in a different way than expected).
- Yields can be improved by changing conditions in reversible reactions to favour products or by recycling unused reactants so they are not wasted.

56. Energy changes

1 **(a)** An exothermic reaction gives out (heat) energy/ transfers energy to the surroundings (1) while an endothermic reaction takes in (heat) energy (1).
(b) Measure temperature change (1); if drop in temperature occurs the reaction is endothermic (1).
(c) Combustion/neutralisation/oxidation of metals (1)

2 **(a)** B: Measure the temperature of each solution (1). C: Add solutions together (stir) and measure the new temperature/find temperature change (1). D: Repeat steps using nitric acid in step A (1).
(b) Any two of: to make the test fair the students would need to use the same volumes of acid and alkali each time (1); the same concentration of acid and alkali each time (1); the same insulated cup or container each time (1).
(c) The one with the greatest temperature rise is the most exothermic (1).

3 It's an endothermic reaction (1) as it is the opposite of combustion (1), which is exothermic. *In an exam you would **not** get credit for saying that photosynthesis needs light energy, as the question asked you to use the given information.*

57. Acids and alkalis

1 **(a)** A and B (1)
(b) E (1)
(c) no change (1)
(d) The student could add some pH (universal) indicator (1) and use the colour chart to find the pH (1).

2 Solutions of orange juice contain excess H^+ ions (1) and solutions of orange juice react with sodium carbonate (1).

3 **(a)** **(i)** H^+ ion (1)
(ii) OH^- ion (1)
(b) **(i)** pH increases (1)
(ii) pH decreases (1)
(c) **(i)** $Na^+(aq)$ (1) and $Cl^-(aq)$ (1)
(ii) $H^+(aq)$ (1) + $OH^-(aq)$ (1) $\rightarrow H_2O(l)$ (1)

58. Making salts

1 **(a)** a salt (1) and water (1)
(b) H^+ (1) and OH^- (1)
(c) increase towards 7 (1)

2 **(a)** base (1)
(b) magnesium chloride (1) and aluminium chloride (1)

3 **(a)** Missing data: lithium chloride (1); nitric acid (1); iron(II) oxide or iron(II) hydroxide (1).
(b) **(i)** $2NH_4OH(aq) + H_2SO_4(aq)$ (1) $\rightarrow (NH_4)_2SO_4(aq) + 2H_2O(l)$ (1)
(ii) fertiliser (1)

4 **(a)** Use an indicator (or pH probe) (1), which changed colour (or showed pH = 7) (1) when the solution was neutralised.
(b) $2HCl$ (1) + $Ca(OH)_2$ (1) $\rightarrow CaCl_2$ (1) + $2H_2O$

59. Making soluble salts

1 **(a)** copper(II) oxide (1) and hydrochloric acid (1)
(b) so all the acid was used up (1)

2 **(a)** $Mg(s) + 2HCl(aq)$ (1) $\rightarrow MgCl_2(aq) + H_2(g)$ (1)
(b) H_2SO_4 (1) + $Zn \rightarrow ZnSO_4 + H_2$ (1)

3 **(a)** Group 1 (1) would be unsuccessful as copper does not react with dilute acid (1).
(b) From top to bottom: filter funnel, filter paper, copper carbonate and salt solution (2 marks for all correct, 1 mark for 3 correct, no marks for only 1 or 2 correct).
(c) The crystals of copper sulfate can be obtained from the salt solution by heating the solution/leaving the solution in an open dish (1) and evaporating the water (1).
(d) **(i)** sodium sulfate (1)
(ii) Sodium is very reactive (1) and it would be dangerous to use (1) (would react violently).

60. Making insoluble salts

1 **(a)** lead nitrate (1)
(b) **(i)** lead nitrate (1) and sodium phosphate (1)
(ii) copper(II) chloride/nitrate/sulfate (1) and sodium carbonate (1)
(c) filtration (1)

2 **(a)** sodium carbonate (1)
(b) $K_2CO_3(aq) + 2AgNO_3(aq) \rightarrow Ag_2CO_3(s)$ (1) + 2 (1) KNO_3 (1) (aq)
(c) as sodium sulfate is soluble in water (1)

3 **(a)** A precipitation forms a solid (1) and the solid contains the ions to be removed from the solution (1).
(b) $2NaOH(aq) + NiCl_2(aq) \rightarrow Ni(OH)_2(s)$ (1) + $2NaCl(aq)$ (1)

61. Using electricity

1 **(a)** bulb would light (1)
(b) Solid lead bromide ions are locked in lattice (1) and cannot move/conduct (1).
(c) The positive electrode product is bromine (1). The negative electrode product is lead (1).
(d) Reduction is the gain of electrons (1) and it occurs at the negative electrode (1).

2 **(a)** The metal ions gain electrons (1) and become metal atoms (1).
(b) oxidation (1)

3 Because the chloride ions are negative, they move towards the positive electrode (1). Because the copper ions are positive they move towards to the negative electrode (1). Oxidation occurs at the positive electrode to form chlorine (1). The oxidation reaction is: $2Cl^-(aq) \rightarrow Cl_2(g) + 2e^-$ (1). Reduction occurs at the negative electrode to form copper (1). The reduction reaction is: $Cu^{2+}(aq) + 2e^- \rightarrow Cu(s)$ (1).

62. Useful substances from electrolysis

1 **(a)** In molten aluminium oxide the ions can move (1). In solid aluminium oxide the ions are locked into the lattice structure (1).
(b) to lower the melting point (1)
(c) Aluminium ions are positive (1) and are attracted to the negative electrode/gain electrons at the negative electrode to form aluminium metal (1).
(d) oxygen (1) and carbon dioxide (1)

2 **(a)** Hydrogen is formed at the negative electrode (1) and chlorine is formed at the positive electrode (1).
(b) $2H^+(aq) + 2e^-$ (1) $\rightarrow H_2$ (1); $2Cl^-$ (1) $\rightarrow Cl_2(g) + 2e^-$ (1)
(c) sodium hydroxide (1), used to make soap (1)
(c) bleach (1) and plastics (1)

63. Electrolysis products

1 **(a)** Missing data: chlorine (1), iodine (1) and hydrogen (1).
(b) A solution that conducts electricity (1) as it contains ions (1) that can move.
(c) They gain electrons (1).
(d) **(i)** Metals near the top of the reactivity series (1).
(ii) hydrogen (1)

2 **(a)** to protect the metal/to improve their appearance (1)
(b) negative (1)
(c) **(i)** electrons (1)
(ii) ions (1)

3 **(a)** $Ag^+(aq) + e^-$ (1) $\rightarrow Ag(s)$ (1)
(b) Electroplating plates a thin coating on the metal (1), and this is an advantage as it uses less of the expensive metal (1).

64. Chemistry six mark question 3

A brief answer would have a simple description of how cobalt chloride can be made from hydrochloric acid and cobalt oxide with some information about separating the excess reactants.

A good answer would have a clear description of how cobalt chloride can be made from hydrochloric acid and cobalt oxide, how you know the reaction is completed and how the cobalt chloride is separated from the excess reactant.

An excellent answer would have a detailed description of how cobalt chloride can be made from hydrochloric acid and cobalt oxide, how you know the reaction is completed, and how you obtain solid cobalt chloride from the solution and excess reactant. There would also be a note of any safety precautions required.

Examples of chemistry points made in the response:

- Using the spatula add excess cobalt oxide to some dilute hydrochloric acid in a beaker and leave for some time (or heat for a few minutes) to react. The reaction is complete if there is excess cobalt oxide and some solid is left at the bottom of the beaker (or a labelled diagram showing this information).
- Use the filter funnel, filter paper and conical flask to filter the mixture to remove unreacted cobalt oxide (or a labelled diagram showing this information).
- Put the solution (filtrate) in an evaporating basin and heat to evaporate water (or leave in a warm place) and collect crystals of cobalt chloride.
- Care is needed in handling acid (and heating solutions). Safety goggles should be worn throughout.

Physics answers

65. Resultant forces

1 **(a)** weight (1)
(b) 100 N (1) upwards (1)

2 **(a)** Total force acting backwards = 2000 N + 4000 N = 6000 N; resultant = 15 000 N − 6000 N = 9000 N (1) to the left/forwards (1)
(b) It will accelerate (1) in the direction of the resultant force (1).

3 **(a)** The force produced by thruster A acts in the direction opposite to the satellite's motion/the resultant force is backwards (1), so the satellite will decelerate/slow down (1) while the thruster is firing. When the thruster stops firing the satellite will continue at its new/slower speed (1).
(b) It will not change the speed (1) because the two forces will cancel each other out/the resultant force will be zero (1).

66. Forces and motion

1 **(a)** $F = m \times a$ = 27 500 kg × 1.5 m/s² (1) = 41 250 (1) N
(b) $a = F/m$ = 1200 N/240 kg (1) = 5 (1) m/s²
(c) $m = F/a$ = 0.6 N/10 m/s² (1) = 0.06 (1) kg

2 **(a)** The balloon is moving horizontally at a constant speed, so there is no resultant force in the horizontal direction (1). The balloon is not moving vertically, so there is no resultant force in the vertical direction (1).
(b) The upthrust/upwards force must have been 30 000 N originally and this has not changed, so the resultant force is now 30 000 N − 29 000 N = 1000 N (1). $a = F/m$ = 1000 N/3000 kg (1) = 0.33 m/s² (1) upwards (1).
(c) There is no change in forces acting in the horizontal direction (1) so the speed in the horizontal direction will not change (1).

67. Distance–time graphs

1 Velocity has a direction as well as a size (1), speed only has a size (1).

2 **(a)** Line drawn from origin to reach 4 km at 4 hours (2). *If you have drawn a line below the one given, starting at the origin but not going to 4 km at 4 hours, give yourself just one mark.*
(b) Line drawn from origin that is steeper than the given one (1), becoming horizontal at 2 km (1).

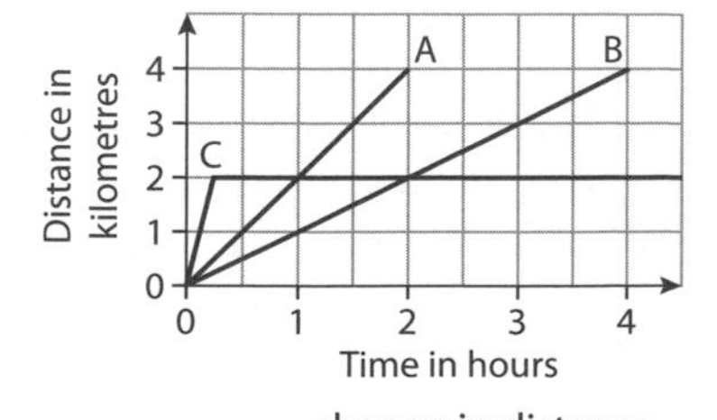

(c) speed = gradient = $\frac{\text{change in distance}}{\text{change in time}} = \frac{4\text{ km}}{2\text{ hours}}$ (1) = 2 (1) km/h

3 **(a)** change in time = 2 minutes = 2 × 60 = 120 seconds (1); gradient = change in distance/change in time = 300 m/120 s (1) = 2.5 (1) m/s
(b) They are moving faster between 7 and 10 minutes (1) because the line is steeper (1). Their velocity is negative/has the opposite sign between 7 and 10 minutes (1) because they are moving in the opposite direction/getting closer to the start instead of further away (1).

68. Acceleration and velocity

1 (a) $a = \frac{v - u}{t} = \frac{14 - 2}{4}$ (1) = 3 (1) m/s²

(b) time = 2 minutes = 2 × 60 = 120 seconds (1); acceleration = (30 − 5)/120 (1) = 0.21 (1) m/s²

2 (a) Lorry B (1) because the line on the graph goes higher (1).

(b) Lorry A (1) because the lines where it is accelerating are steeper (1).

3 (a) acceleration = gradient = $\frac{\text{change in velocity}}{\text{change in time}}$

(20 − 0)/(30 − 10) = 20/20 (1) = 1 (1) m/s²

(b) acceleration = (10 − 20)/(70 − 50) = −10/20 (1) = −0.5 (1) m/s² *You must have the minus sign to get the 2nd mark.*

(c) distance travelled = area under graph = area of triangle + area of rectangle = (0.5 × 20 s × 20 m/s) + (20 s × 20 m/s) (1) = 200 + 400 = 600 (1) m

69. Forces and braking

1 (a) As speed increases the braking distance increases (1) in a non-linear way. *Non-linear means that a graph of braking distance against speed would not be a straight line (equal increases in speed do not give equal increases in braking distance).* The distance increases faster than the speed (1).

(b) A greater force would be needed (1).

(c) The friction between the brakes and the wheels reduces the kinetic energy of the vehicle (1). This energy is transferred to the brakes, which become hotter (1).

2 (a) The distance a vehicle travels while the driver is reacting to a hazard ahead/during the driver's reaction time (1).

(b) Any two from: tired driver (1); driver on drugs (1); driver has been drinking alcohol (1).

(c) adverse road conditions/icy road/wet road (1); poor brakes (1)

3 (a) The thinking distance is the distance the car travels while the driver is reacting to a hazard/obstacle ahead (1), and so depends only on his/her reaction time and not on the vehicle (1).

(b) Any two from: the braking distances given assume the brakes are applied fully and the car does not skid (1). If the driver is not alert/concentrating, the driver may not press the brakes hard enough (1) or may lose control of the car on wet/icy roads (1).

70. Falling objects

1 (a) $W = m \times g$ = 76 kg × 10 N/kg (1) = 760 (1) N

(b) mass in kg = 80 g/1000 = 0.08 kg (1); $W = m \times g$ = 0.08 kg × 10 N/kg (1) = 0.8 (1) N

2 (a) The drag force/water resistance increases as the speed increases (1).

(b) $m = W/g$ = 200 000 000 kg/10 N/kg (1) = 20 000 000 (1) kg

3 (a) Weight does not change (1).

(b) Any four from: air resistance increases initially as her speed increases (1), it then remains constant while she is falling at terminal velocity (1). When she opens her parachute the air resistance increases again (1) because the area has increased (1). As she slows down her air resistance decreases (1) and then it remains constant when she is falling at her new terminal velocity (1).

71. Forces and terminal velocity

1 Labels as shown below (1 mark for each correct)

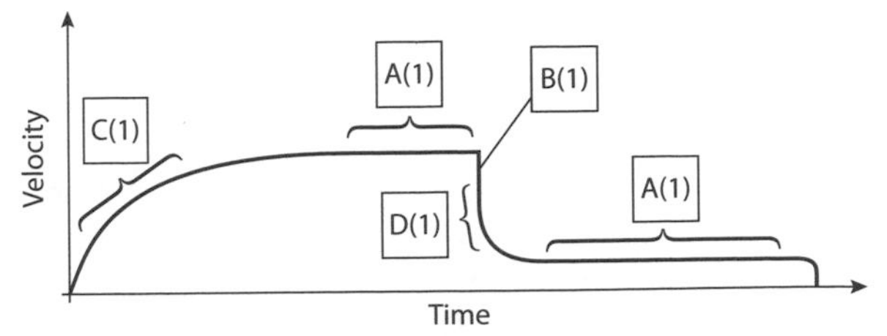

2 (a) (i) Sketch graph showing velocity increasing to 400 km/h at 4 seconds (1) with the line steep at first and then getting shallower (1), then a horizontal line to 7 seconds (1) and then a steep downwards line (1). *The question did not say how long it took for the parachute to slow the car down, so it doesn't matter whether you showed the car reaching a stop.*

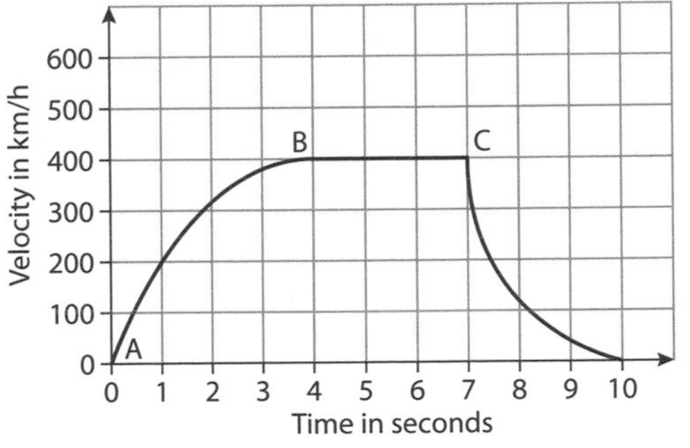

(ii) Any four from: when the car is just starting the air resistance is zero/very small and so the resultant force is large (1), the acceleration is large (1) and the gradient is steep (1). As the car speeds up the air resistance increases (1), so the resultant force reduces (1) and the acceleration gets less (1), so the gradient of the velocity–time graph also gets less (1).

(b) Letters marked as shown above (1 for each correct placement).

72. Elasticity

1 (a) $F = k \times e$ = 3.5 N/m × 4 m (1) = 14 (1) N.

(b) 20 cm = 20/100 m = 0.2 m (1); $F = k \times e$ = 120 N/m × 0.2 m (1) = 24 (1) N.

2 Extension = 7 cm − 5 cm = 2 cm (1), 2 cm = 0.02 m (1); OR extension = 0.07 m − 0.05 m (1) = 0.02 m (1); $k = F/e$ = 20 N/0.02 m (1) = 1000 (1) N/m

3 Any four from: the work done as the force pulls the toy down (1) is stored as elastic potential energy (1); this is converted to kinetic energy as the toy moves up (1) and then to gravitational potential energy as the toy goes above its original position (1). This is then converted back to kinetic energy and to elastic potential energy as the toy goes down again (1).

73. Forces and energy

1 Force B and distance X (1), as force B is the force being used to push the barrow (1) and distance X is the distance in the direction of the force (1).

2 $W = F \times d$ = 50 N × 5 m (1) = 250 (1) J (1)

3 $P = E/t$ = 20 J/2 s (1) = 10 (1) W (or watts) (1)

4 Distance = 50 cm = 0.5 m (1); $F = W/d$ = 150 J/0.5 m (1) = 300 N (1)

5 *First calculate the energy transferred (the work).* $W = F \times d$ = 8000 N × 25 m (1) = 200 000 (1) N; $P = E/t$ = 200 000 N/4 s (1) = 50 000 (1) W

74. KE and GPE

1 $E_p = m \times g \times h = 0.8\text{ kg} \times 10\text{ N/m} \times 1.5\text{ m}$ (1) = 12 (1) J
2 180 g = 0.18 kg (1) $E_p = 0.18\text{ kg} \times 10\text{ N/kg} \times 4\text{ m}$ (1) = 7.2 (1) J
3 $E_k = \frac{1}{2} \times m \times v \times v = 0.5 \times 4\text{ kg} \times 1.5\text{ m/s} \times 1.5\text{ m/s}$ (1) = 4.5 (1) J
4 Any three from: the crumple zone collapses when the car hits something (1) and it needs energy/work to do this (1), so this transfers the kinetic energy of the car into other forms of energy (1), so less energy is transferred to the occupants (1).
5 E_p of the book on the shelf $= m \times g \times h$ $= 1\text{ kg} \times 10\text{ N/kg} \times 2\text{ m}$ (1) = 20 J (1). This is transferred to kinetic energy, so kinetic energy when it hits the floor is 20 J (1). $v^2 = 2E_k/m = (2 \times 20\text{ J})/1\text{ kg} = 40$ J (1); $v = 6.3$ (1) m/s.

75. Momentum

1 160 g = 0.16 kg (1); $p = m \times v = 0.16\text{ kg} \times 2\text{ m/s}$ (1) = 0.32 (1) kg m/s
2 If the direction is reversed, the velocity has the opposite sign (1) so the momentum would have the opposite sign/be a negative value (1).
3 **(a)** $p = m \times v$, total momentum = (1.5 kg × 0.2 m/s) + (1 kg × 0.1 m/s) (1) = 0.4 kg (1) m/s
(b) Total momentum after the collision will be 0.4 kg m/s (1); $v = p/m = 0.4\text{ kg m/s}/2.5\text{ kg}$ (1) = 0.16 (1) m/s. *If you got the answer to part (a) wrong, but used that answer correctly in part (b), you can give yourself the marks.*
4 **(a)** If the bomb was not moving it had zero momentum before it exploded (1). If pieces fly off in opposite directions, their momentum will have opposite signs/some will have positive and some will have negative momentum (1). The sum of the momentum of all the pieces adds up to zero (1).
(b) $p = m \times v$, momentum of 4 kg piece = 4 kg × 240 m/s (1) = 960 kg m/s (1); momentum of 6 kg piece = −960 kg m/s (1) *You must have the minus sign to get this mark.* $v = p/m = -960\text{ kg m/s}/6\text{ kg}$ (1) = −160 (1) m/s *You must have the minus sign to get this mark.*

76. Physics six mark question 1

A basic answer will include some mention of a safety feature or safety features, or some scientific ideas about energy changes in crash situations.

A good answer will include some explanation of how kinetic energy is transferred to other forms of energy in normal braking or in a crash and how safety features help to protect occupants of a car.

An excellent answer will include a clear, balanced and detailed explanation of the conversion of kinetic energy to other forms of energy, and the function of air bags, crumple zones and seat belts.

Examples of physics points made in the response:

- During normal slowing/stopping, the brakes convert the kinetic energy of the car into other forms of energy, which end up heating the brakes and the surroundings.
- In a crash, work is done to collapse crumple zones and air bags and to stretch seat belts.
- Side impact bars spread some of the force to stronger parts of the car, so that the occupants are not crushed by a side impact.
- These safety features help to bring the occupants of the car to a more gradual halt.
- If the occupants are not wearing their seat belts, they would carry on moving forwards when the car stops, and then stop very suddenly (causing severe injuries) when they hit the windscreen/steering wheel.

Although it is not required in the specification, you may also gain credit for discussing anti-lock braking systems (ABS), which help to stop a car skidding. You will not gain credit for discussing regenerative braking systems, as these are intended to save energy during normal driving and do not contribute to the safety of a vehicle.

77. Static electricity

1 **(a)** Electrons were transferred (1) from the cloth to the rod (1).
(b) They will move apart (1) because like charges/the same type of charges repel (1).
(c) Protons are in the nucleus of atoms and cannot move (1). The rod has a positive charge because it has lost electrons (1).
2 Combing can give the hair a charge of static electricity (1) by transferring electrons to/from the comb (1). All the strands of hair will have the same charge (1) so they repel each other (1).
3 Student B is correct. Any three from: when electrons are transferred to a metal they can spread out through the metal (1) so a charge does not build up (1) and it looks as if there has been no charge transferred (1). When an insulating material is rubbed, the electrons transferred cannot move about (1).

78. Current and potential difference

1 **(a)** $I = \frac{Q}{t}$ = 600 C/40 s (1) = 15 (1) A
(b) $V = \frac{W}{Q}$ = 240 000 J/600 J (1) = 400 (1) V
2 **(a)** $Q = I \times t$; charge from 0 to 4 minutes = 8 A × 4 × 60 s (1) = 1920 C (1); charge from 4 to 6 minutes = 4 A × 2 × 60 s = 480 C (1); total charge = 1920 C + 480 C = 2400 (1) C
(b) $E = V \times Q$ = 230 V × 2400 C (1) = 552 000 (1) J *If you got the wrong answer for part (a), but you used that number correctly in part (b), you can get both the marks for part (b).*
(c) **(i)** $Q = E/V$ = 200 000 J/110 V (1) = 1818 C (1)
(ii) $I = \frac{Q}{t}$ = 1818 C/(6 × 60 s) (1) = 5.05 A (1).
If you get the wrong answer for part (i), but you used the number correctly in part (ii), you can still get both marks for part (ii).

79. Circuit diagrams

1 Circuit A: correct symbols for battery (1), resistors (1), arranged as shown (1).

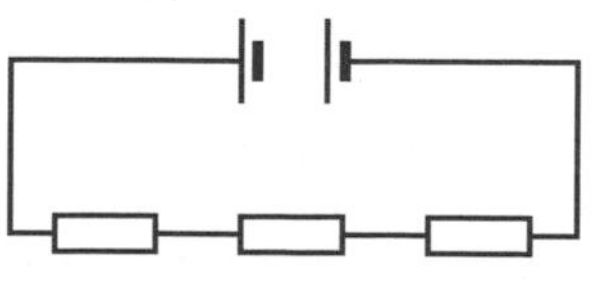

Circuit B: correct symbols for ammeter (1), variable resistor (1), arranged as shown (1).

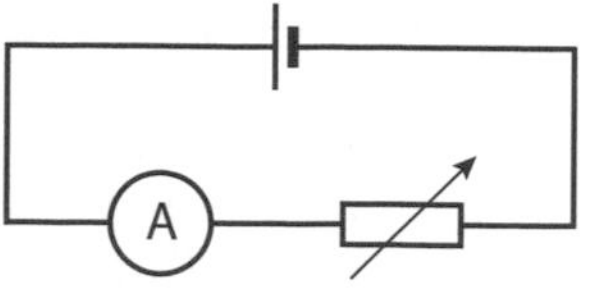

2 Correct symbol for cell (1), resistor (1), lamp (1), correctly drawn in circuit with no gaps, as shown (1). *It doesn't matter if you have put the resistor and bulbs in slightly different places, as long as the bulbs are in parallel to each other and the resistor is in series.*

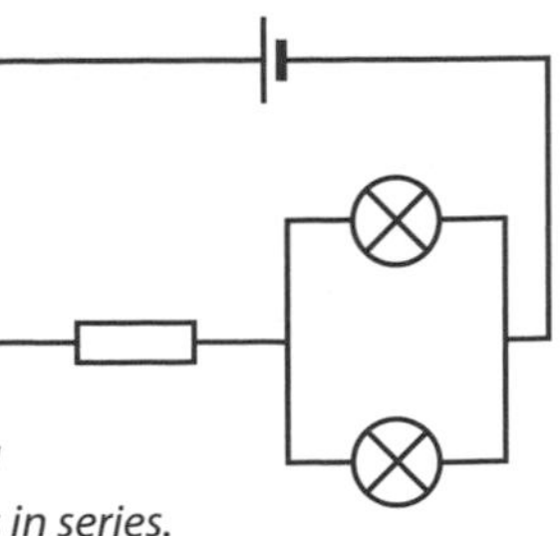

3 A – thermistor (1), B – LED (1), C – diode (1), D – LDR (1)

4 Correct symbol for ammeter and voltmeter (1), arranged as shown (1).

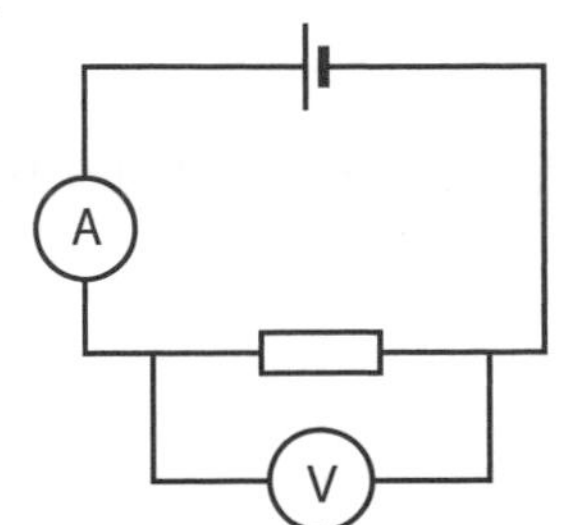

80. Resistors

1 (a) For a given voltage/potential difference, the current is lower in A than in B (1), so it must have a higher resistance (1). *You also get full credit if you explain in terms of B having more current for a given potential difference than A, so it must have a lower resistance.*

(b) 0.25 A (1)

2 $V = I \times R = 1.5\ \text{A} \times 800\ \Omega$ (1) = 1200 (1) V

3 (a) $I = V/R = 12\,000\ \text{V}/1000\ \Omega$ (1) = 12 (1) A

(b) $R = V/I = 12\ \text{V}/0.01\ \text{A}$ (1) = 1200 Ω (1)

4 Resistor A: current at 5 V is 0.5 A. $R = V/I = 5\ \text{V}/0.5\ \text{A}$ (1) = 10 (1) Ω; resistor B: current at 6 V = 1.5 A. $R = V/I = 6\ \text{V}/1.5\ \text{A}$ (1) = 4 (1) Ω. *You can use any pairs of values from the lines on the graph – you should get the same answers.*

81. Series and parallel

1 (a) 4.5 V (1); **(b)** 0 V (1); **(c)** 1.5 V (1)

2 (a) (i) 0.4 A (1); **(ii)** 4 V (1); **(iii)** 12 V (1); **(iv)** 12 V (1); **(v)** 1.5 A (1)

(b) resistance = 5 Ω + 10 Ω (1) = 15 (1) Ω

3 (a) total resistance in that branch = 12 Ω (1); potential difference across both resistors = 6 V (1); $I = V/R = 6\ \text{V}/12\ \Omega$ (1) = 0.5 (1) A

(b) Both will read 3 V (1), as there is a total of 6 V across that branch of the circuit (1) and as both have the same resistance the potential difference will be divided between them equally (1).

(c) *There are two different ways you can work out the answer to this question. Either method is valid. Ratios:* The ratio of the two resistances is 1 : 5 and so the potential differences across the two resistors will be in the ratio 1 : 5 (1). The potential difference across both is 6 V (1), so V1 = 1 V and V2 = 5 V (1). OR *Calculating using $V = I \times R$:* Current through both resistors is 0.5 A (1). For 2 Ω resistor, $V = 0.5\ \text{A} \times 2\ \Omega$ = 1 V (1). For 10 Ω resistor, $V = 0.5\ \text{A} \times 10\ \Omega = 5\ \text{V}$ (1).

82. Variable resistance

1 (a) LDR (or light-dependent resistor) (1)

(b) The resistance gets less (1) when brighter light shines on it (1). *It is not enough to say that the resistance changes – you need to say how it changes, i.e. the resistance decreases.*

2 (a) diode or LED/light-emitting diode (1)

(b) Line at zero or just above zero on the left hand side of the graph (1), sloping line on right hand side (1); change from zero to sloping line just to the right of the vertical axis (1).

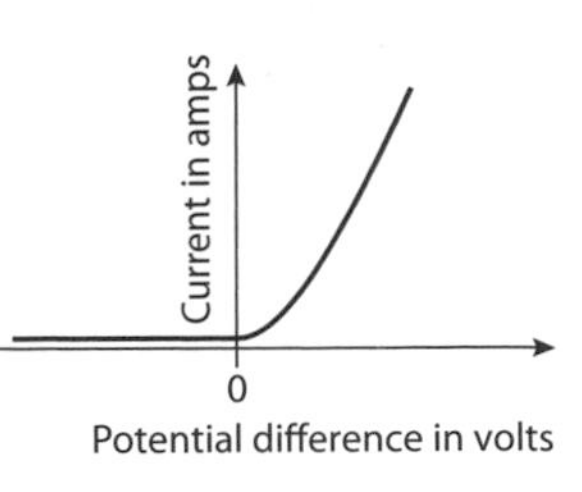

3 (a) make it colder (1)

(b) Any sensible use that involves changing something when the temperature changes, for 1 mark. Examples include switching on a warning light or buzzer if the inside of a fridge gets too warm, or as a thermostat to control the temperature in a house or in an oven.

4 Any four from: the first part of the line is straight because the resistance does not change (1) and the current increases as the potential difference increases (1). At higher currents the filament gets hotter and the ions in the metal vibrate more (1); the vibrating ions make it harder for the electrons to flow between them (1), so the resistance increases (1) and each increase in potential difference only produces a smaller increase in current (1).

83. Using electrical circuits

1 (a) Filament bulbs use a lot more electrical power/energy than other types of bulb for the same light output (1). Most electricity is generated using fossil fuels (1) and burning these is contributing to global warming/climate change (1).

(b) LED bulbs are better (1). Plus any three points from the following: LED bulbs use less power that compact bulbs for the same light output (1) and they last longer (1). They cost a lot more/2.9× more (1), but the extra life makes up for this/but they last more than 6 times as long (1). LED bulbs are easier to dispose of (1).

2 (a) The thermistor is taking most of the potential difference from the cell (1) so the potential difference across the LED is not enough to allow it to conduct/light up (1).

(b) The LED will start to shine when the temperature gets high enough (1), as the resistance of the thermistor decreases with temperature (1) and so it will take less of the potential difference/the LED will get a higher proportion of the potential difference (1).

84. Different currents

1 (a) In a direct current, the current always flows in the same direction (1). An alternating current is constantly changing direction (1).

(b) 230 (1) V, 50 (1) Hz

(c) battery/cell (1)

2 (a) 2 (1)

(b) 4 V (1)

(c) Horizontal straight line (1) drawn 3 squares up (1).

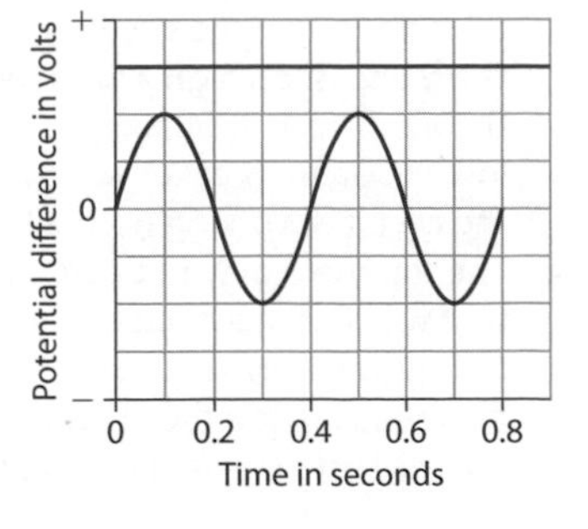

3 (a) 0.4 seconds (1)
(b) frequency = 1/period = 1/0.4 s (1) = 2.5 (1) Hz
(c) period = 1/frequency = 1/30 Hz (1) = 0.033 (1) s
(d) One complete cycle shown, symmetrical about the horizontal zero axis (1) with a period of 0.5 s (1) and peak voltage of 3 V (1), as shown.

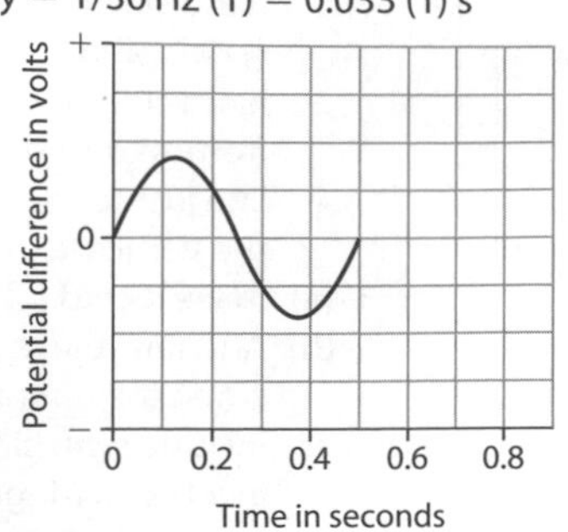

85. Three-pin plugs

1 (a) A three-core cable consists of three metal wires (1), each wire is surrounded by a layer of (coloured) plastic (1), and all three wires are held together by an outer layer of plastic (1).
(b) (i) Copper is a good conductor (of electricity) (1).
(ii) Plastic is an insulating material (1), and the cable needs to bend so the plastic must be flexible (1).
2 (a) live – brown (1); neutral – blue (1); earth – green and yellow (1) *You need to give both colours for the earth wire.*
(b) earth (1)
(c) Each wire does a different job (or similar explanation) (1), so the colours help to make sure the correct wires are connected (1).
3 Any four from: brass is used for the pins because it is stronger than copper, but is not such a good conductor of electricity (1). Strength is more important than its conducting abilities for the pins (1), as being strong will prevent the pins being damaged as they are put in and taken out of sockets (1). And any two from: a three-core cable consists of three metal wires (1). Copper is used for the wires as it is a better conductor than brass (1) and there is a much greater length of wire than pin in a cable and plug (1). Strength is not so important for the wire inside the cable (1).

86. Electrical safety

1 (a) If the current gets too high (1), the fuse melts (1) and breaks the circuit/cuts off the electricity (1).
(b) It protects the wiring/can prevent fires (1).
(c) (i) It compares the currents in the live and neutral wires (1), and if they are different it cuts off the electricity (1).
(ii) It works faster (1).
2 (a) It has a casing made out of an insulating material (1).
(b) It only has two wires/it has no earth wire (1).
3 (a) So a fault in one circuit does not remove power from the whole house/home (1).
(b) Any three from: different circuits will have different currents/lighting circuits will have lower currents than circuits for sockets/cookers/showers (1). The current rating should be just above the maximum current expected when the circuit is working correctly (1). If the rating is too high a fire/overheating could occur before the circuit breaker operates (1), and if it is too low the circuit breaker will cut off the electricity when the circuit is working normally (1).
(c) The thickness of wire depends on the current that will flow in a circuit (1). Higher currents need thicker wires (1). If the wire is too thin the wire may overheat/cause a fire (1). It would cost more to use the thickest wire needed for all the wiring (1).

87. Current and power

1 1 hour = 60 × 60 = 3600 seconds (1);
$P = \frac{E}{t}$ = 72 000 J/3600 s (1) = 20 (1) W
2 $P = I \times V$ = 1.5 A × 3 V (1) = 4.5 (1) W
3 Less energy is wasted as heat (1).
4 (a) $P = \frac{E}{t}$ = 120 J/0.01 s (1) = 12 000 W (1)
(b) $I = P/V$ = 12 000 W/1500 V (1) = 8 (1) A
(c) $Q = E/V$ = 120 J/1500 V (1) = 0.08 (1) C
OR $Q = I \times t$ = 8 A × 0.01 s (1) = 0.08 (1) C

88. Physics six mark question 2

A basic answer will include a brief reference to charges being transferred, and to the two polythene rods repelling each other or the polythene and Perspex rods attracting each other.
A good answer will include an explanation of electrons being transferred, and make reference to like charges repelling and unlike ones attracting, and to charges moving easily through metals.
An excellent answer will include a clear, balanced and detailed explanation of how electrons are transferred between rod and cloth to produce positive and negative charges, why a metal rod will not show a static charge, and what will be seen when different pairs of rods are suspended next to each other.

Examples of physics points made in the response:

- Rubbing can transfer electrons from one material to another.
- The material that gains electrons will have a negative charge, the material that loses electrons will have a positive charge.
- When polythene is rubbed, electrons are transferred from the cloth to the rod, giving the rod a negative charge and leaving the cloth with a positive charge.
- When Perspex is rubbed, electrons are transferred from the rod to the cloth, leaving the rod with a positive charge and giving the cloth with a negative charge.
- When a metal rod is rubbed, the electrons can move freely through it so they spread out and we do not notice a charge.
- If the two polythene rods are suspended next to each other they will repel, because they both have the same type of charge.
- If a polythene and Perspex rod are suspended next to each other they will attract, because they have opposite charges.
- The metal rod will not affect the other rods.

89. Atomic structure

1 (a) 11 (1), 21 (1)
(b) Particles A and D (1) are isotopes of the same element because they have the same number of protons/atomic numbers (1) but different numbers of neutrons (1).
(c) (i) Particles C and D are ions (1) because they do not have the same number of protons and electrons (1).
(ii) C is a positive ion (1) as it has more protons than electrons/it has lost an electron (1).
2 proton: charge = +1 (1), position = nucleus (1);
neutron: mass = 1 (1), position = nucleus (1);
electron: charge = −1 (1)

3 The atomic number is 17, so a chlorine atom has 17 protons (1). The mass number is not a whole number, so there must be isotopes of chlorine with different numbers of neutrons (1). Some of these isotopes will have 18 or fewer neutrons, and some will have 19 or more (1).

90. Background radiation

1 **(a)** nucleus (1)
(b) It will not change (1).
(c) It is not possible to predict when a particular nucleus will decay/give out radiation (1).
2 true activity = measured activity − background count = 340 − 24 (1) = 316 (1)
3 There may be more natural radiation present from rocks/buildings/radon (1); there may be more radiation from human activities such as nuclear power or hospitals (1).
4 **(a)** The section showing radiation from the ground will be larger (1).
(b) The pie charts only show the proportions/percentages of the background radiation that comes from different sources (1); they do not show the total background radiation (1).

91. Nuclear reactions

1 **(a)** alpha: +2 (1), 4 (1);
beta: electron (1), 0 (1)
(b) A helium nucleus has two protons and two neutrons (1), which is the same structure as an alpha particle (1).
(c) $^{226}_{88}Ra$ (1 mark for each correct number)
2 **(a)** $^{239}_{94}Pu \rightarrow ^{235}_{92}U + ^{4}_{2}He$
(1 mark for each correct number)
(b) $^{99}_{43}Tc \rightarrow ^{99}_{44}Ru + ^{0}_{-1}e$
(1 mark for each correct number)
3 α (1), α (1), β (1)

92. Alpha, beta and gamma radiation

1 **(a)** alpha (1) **(b)** gamma (1) **(c)** beta (1)
2 Alpha particles are not very penetrating (1), so they cannot get into the body if the source is outside (1). They are very ionising (1), so they can cause a lot of damage if they are emitted inside the body (1).
3 **(a)** They have opposite charges (1).
(b) There is a greater force on alpha particles because they have more charge (1), but they also have a much greater mass (1), which results in them being deflected less than beta particles (1). *You could also have answered in terms of beta particles having a smaller force because they have a smaller charge, but being deflected more because they have a smaller mass.*
(c) It would have no effect (1) because gamma radiation does not have a charge/is an electromagnetic wave (1).

93. Half-life

1 Working shown (this can be lines on the graph, or notes of times and count rates) (1), half-life = 6.5 minutes (1). For example, the count rate is 600 per second at 5 minutes. It is half of this (300) at 11.5 minutes. Half-life = 11.5 − 5 minutes (1) = 6.5 minutes (1).
2 30 minutes is 3 half-lives (1); 1000 → 500 → 250 → 125 (1); activity = 125 counts per minute (1)
3 **(a)** A radioactive substance is injected into a patient (1) and travels to cancer cells (1) where it collects and radiation from the tracer is detected using a gamma camera/special camera (1).
(b) Tc-99m is used (1). Any three from: because it emits gamma radiation (1), which is the only form of radiation that can pass through the body without causing ionisation/damage (1). It also has a short half-life (1) so the radiation inside the patient does not last very long (1).

94. Uses and dangers

1 **(a)** It can ionise atoms (1), which can lead to diseases such as cancer (1).
(b) Any three from: wear masks when using alpha sources (1), point sources away from the body (1), handle sources with tongs (1), keep sources in lead-lined boxes (1).
2 The radiation needs to be able to travel through the earth above the pipe to reach the detector (1). Gamma radiation is the only one with enough penetrating ability to do this (1).
3 **(a)** Any four from: the paper absorbs some of the radiation from the source (1). If the paper is too thick, more radiation is absorbed/less gets through to the detector (1) and the processer increases the pressure on the rollers (1). If the paper is too thin less radiation is absorbed/more gets through (1) and the processor reduces the pressure on the rollers (1).
(b) Beta radiation is used (1). Alpha radiation would not penetrate the paper at all (1) and gamma would penetrate too easily so the amount of radiation would not vary much with the thickness of the paper (1).

95. The nuclear model of the atom

1 **(a)** The atom is mostly empty space (1).
(b) The nucleus has the same charge as alpha particles (1).
2 New evidence was obtained (1), which could not be explained using the previous model (1).
3 Both models included positive and negative charges (1). In the plum pudding model the positive charge is spread throughout the atom, but in the nuclear model it is concentrated in the nucleus (1). In the plum pudding model the mass is spread out over the whole volume of the atom, but in the nuclear model it is concentrated in the nucleus (1).
4 **(a)** Alpha particles have positive charges, and so will only be repelled by positive charges (1). If the positive charge was spread out through the atom it would not be concentrated enough to make the alpha particles bounce back (1).
(b) Only a few particles had been reflected (1), so the mass must occupy only a small part of the atom (1).

96. Nuclear fission

1 uranium-235 (1), plutonium-239 (1)
2 **(a)** 3 neutrons drawn at the ends of the 3 central arrows (1). Labels (left to right): neutron (1), uranium/plutonium nucleus (1), daughter nucleus/smaller nucleus (1), neutrons (1).
(b) The neutrons released by the fission reaction (1) can all hit other nuclei and make them split up (1), and when these split up, even more neutrons are produced (1).
3 **(a)** Any two from: water is heated to produce steam (1), the steam drives turbines (1) and the turbines drive the generators (1).

(b) A gas-fired power station burns gas to produce heat, a nuclear power station uses heat released in fission reactions (1).

4 Similarity: both result in new isotopes/elements being formed (1). Differences: radioactive decay is a random process, but fission happens when a nucleus is struck by a neutron (1), radioactive decay results in alpha, beta or gamma radiation being emitted, but fission results in neutrons being emitted (1).

97. Nuclear fusion

1 Both produce new elements (1). In nuclear fission this is done by splitting up a large nucleus (1) but in fusion it happens when two small nuclei fuse to make a bigger one (1).

2 **(a)** hydrogen (1)

(b) **(i)** He, C, Fe (2 marks for all three, 1 mark if only one or two)

(ii) in supernova explosions (1)

3 Hydrogen was formed in the Big Bang (1), all the other elements were made inside stars or in supernova explosions (1) and spread into space when earlier stars exploded (1).

4 Fusion reactions only take place at very high temperatures and pressures (1). These conditions are very difficult to achieve on Earth (1) and we do not yet have the technology to do this (1). Fission reactions can be made to happen much more easily (1).

98. The lifecycle of stars

1 **(a)** red giant (1)

(b) white dwarf (1)

(c) supernova (1)

(d) black hole (1)

2 **(a)** A (1)

(b) They have more mass/are more massive (1). *It is not enough to say that the stars are bigger, you have to say that they have more mass.*

(c) They have different masses (1); stars that become black holes (*or whatever you put in answer to question 1(d), if you got that wrong*) are more massive than stars that become neutron stars (1).

3 **(a)** Dust and gas in space (1) were pulled together by gravitational attraction (1) and conditions inside became hot enough/high enough pressure for fusion reactions to start (1). Smaller accumulations of dust/gas became the planets (1).

(b) main sequence (1)

(c) **(i)** remains in the same state/does not change (1)

(ii) The forces within it are balanced (1).

99. Physics six mark question 3

A basic answer will include a brief reference to how a smoke alarm works or how a medical tracer is used, and some mention of the penetrating properties of different forms of radiation.

A good answer will include an explanation of how a smoke alarm works or how a medical tracer is used, with an explanation of the most suitable isotope in terms of penetrating ability, ionising ability and half-life.

An excellent answer will include a clear, balanced and detailed explanation of how radioisotopes are used in both applications, with an explanation of the most suitable isotope for each application in terms of penetrating ability, ionising ability and half-life.

Examples of physics points made in the response:

- Medical tracers identify problems within the body.
- The tracer needs to emit gamma rays as these are penetrating/can pass through the body.
- Gamma rays are also suitable as they are not very ionising and so will not cause much damage to the body.
- The isotope should have a short half-life, so the level of radiation within the patient reduces quickly, so technetium-99m is the best one to use.
- The half-life of barium-137m is too short to be useful.
- Smoke alarms work by ionising the air inside to allow a small current to flow; smoke stops this current.
- An alpha emitter is suitable as alpha particles are highly ionising.
- Alpha particles are also suitable because they are not penetrating and will not travel far in air, so will not be a danger to people.
- A long half-life is necessary, so that the radiation does not reduce much during the life of the smoke alarm, so americium-241 is the best one to use.

Practice paper answers

Additional Science Biology B2 practice paper

1 **(a)** **(i)** chloroplast (1)

(ii) absorb sunlight (1); for photosynthesis/to make sugars/glucose food (1)

(b) label to either the cell wall or the vacuole (1)

(c) more chloroplasts (1)

2 A basic answer will give limited information about some simple ways to collect good data.

A good answer will give simple information on how methods of collecting valid/reliable data should form part of the experiment.

An excellent answer will give a complete description of ways of collecting valid and reliable data, including repeats and removing anomalies.

Examples of biology points made in the response:

- Use of quadrats to sample areas/use a transect perpendicular to the road.
- Repeating for several quadrats at the same distance/repeat with transects at different places along the road.
- Finding a mean number of plants for each distance.
- Method for assessing for and accounting for anomalous results.
- Ensuring quadrat is the same/same size for each trial.
- Stretch of road with same traffic flow.

3 **(a)** Heart rate rises very rapidly initially (1); then levels out at a maximum value (1).

(b) Any three from: faster blood flow to the muscles (1); more oxygen delivered to muscles (1); to make sure that respiration in the muscles continues (1); which releases energy (1).

(c) **(i)** anaerobic respiration (1)

(ii) muscle cells become fatigued (1); and will stop contracting efficiently (1)

4 **(a)** isolation (1)

(b) Not extinct because extinction means a species completely dies out (1); although lemurs died out in Africa, they are still found in Madagascar (1).

(c) Any two from: monkeys passed on diseases to lemurs (1); monkeys killed the lemurs (1); monkeys were more successful in competing for resources/habitat (1).

(d) Fossil record is incomplete/many parts of the organisms will decay and not form fossils (1).

5 (a) sperm and egg/ovum (1)

(b) meiosis (1); cell divides twice/to form 4 daughter cells (1); each daughter cells having a single set of chromosomes/23 chromosomes (1)

(c) (i) correct gametes for parents, i.e. X and X for females and X and Y for males (1); correct genotypes for offspring (1); explanation that there are 2 male and 2 female possible phenotypes (1)

			Father: **XY**	
			gametes	
			X	Y
Mother: **XX**	gametes:	X	XX	XY
		X	XX	XY

(ii) Chance is random/females live longer than males (1). *Remember that a 'suggest' question asks you to apply your knowledge to a new scenario, so there may be other sensible suggestions that the examiners would credit.*

6 (a) The condition only develops if the offspring inherit two identical copies (1); of the allele for the condition (1).

(b) both parents having **Aa** as genotype (1); correct gametes shown for both parents (1); offspring correct in genetic diagram (1); and **aa** indicated as the phenotype for albinism (1)

			Father: **Aa**	
			gametes	
			A	a
Mother: **Aa**	gametes:	A	AA	Aa
		a	Aa	aa

(c) Any three from: the mice best suited to each climate would be most likely to survive/the white mice would be more likely to survive in the snowy island (1); the frequency of each type of mouse coat colour would change in the population (1); over time the two populations could become very different (1); until eventually they are too different to breed together successfully and they become different species (1).

7 (a) Converts glucose (syrup) into fructose (syrup) (1); which is much sweeter (1); and can be used in foods for slimmers (1).

(b) Any three from: enzyme activity increases with temperature (1); until the optimum temperature/about 60 °C is reached (1); after this the activity drops sharply (1); as the enzyme changes shape/is denatured, which stops it functioning (1).

8 (a) structure containing different types of tissues working together (1)

(b) bone marrow (1)

(c) You need at least one argument in favour and at least one argument against (total of 3 marks). The final mark is for a conclusion, showing how you balance the arguments. Arguments in favour: stem cells can turn into many types of cell (1); large numbers of possible treatments (1). Arguments against: ethical concerns about the use of embryonic stem cells (1); treatments likely to be expensive (1). Conclusion: for example, despite the possibilities, it is difficult to justify because of the ethical problem of destroying human embryos to collect the stem cells/even though the research will be expensive and take time, the potential benefits are so major that it is justified (1).

9 (a) to produce fat/oil/starch (1); for energy storage (1) OR to produce cellulose (1); to strengthen the cell wall (1)

(b) Increasing light intensity increases the rate of photosynthesis (1); rate of photosynthesis reaches a maximum (although light intensity keeps increasing) (1); therefore, some other factor must be limiting the rate of photosynthesis (1); which is likely to be temperature/carbon dioxide concentration (1).

Additional Science Chemistry C2 practice paper

1 (a) ionic bonding (1) and lattice structure (1)

(b) Electrostatic forces (1) are attractions between the opposite charges that hold (the positive and negative) ions together (1).

(c) No (1) because the ions are not free to move (1).

2 (a) (i) correct electron arrangement (1); 2^- sign (1)

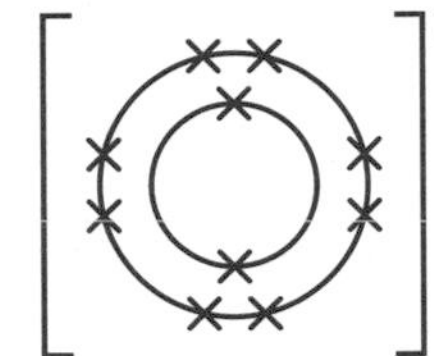

(ii) $2Ca$ (1) + O_2 (1) $\rightarrow$ $2CaO$

(b) (i) $CaCl_2$ (1)

(ii) 2^+ (1)

(iii) fluorine (1)

(c) argon (1)

3 (a) They are very small (1).

(b) Nanoparticles are so small they can get into every cell/full properties not known (1).

(c) Graphite/diamond (1) is giant/lattice structure/macromolecule (1).

(d) Catalysts speed up a reaction without being used up (1). Therefore they only need to be bought once/can be used over and over again (1).

4 (a) Missing labels: hydrogen (at the negative electrode) (1) and sodium chloride (1).

(b) (i) $2Cl^-(aq)$ (1) $\rightarrow$ $Cl_2(g) + 2e^-$ (1)

(ii) oxidation (1)

(c) The reaction produce H^+ ions (1), which make the solution acidic (1).

5 (a) $S(s) + O_2(g)$ (1) $\rightarrow SO_2(g)$

(b) The reaction is reversible/goes both ways (1).

(c) (i) The higher the temperature the lower the % conversion to SO_3 (1).

(ii) 70% conversion (1); 8 tonnes SO_2 should give 10 tonnes SO_3; 70% = $10 \times 70/100$ (1) = 7 tonnes (1)

(d) 1 mole $H_2SO_4 \rightarrow$ 1 mole SO_3; 98 g $\rightarrow$ 80 g; 1 g = 80/98 = 0.82; 400 g = 400×0.82 (1) = 328 g (1)

6 A basic answer would give a simple description of how temperature, concentration and particle size affect the rate of reaction between marble chips and hydrochloric acid. A good answer would give a clear explanation of how temperature, concentration and particle size affect the rate of reaction between marble chips and hydrochloric acid with some reference to how reactions occur.

An excellent answer would give a detailed explanation of how temperature, concentration and particle size affect the rate of reaction between marble chips and hydrochloric acid with careful reference to the collision theory and activation energy.

Examples of chemistry points made in the response:

- Reaction occurs when reacting molecules collide with sufficient energy
- called the activation energy.
- Reaction rates increase with increasing temperature
- as more collisions occur and more molecules have activation energy.
- Reaction rates increase with increasing concentration
- as molecules closer together so more collisions.
- Reaction rates increase with decreasing particle size
- as smaller particle have larger surface area and therefore more collisions occur.

7 (a) It stops fizzing/some powder is left (1).

(b) The unreacted cobalt carbonate is solid (1) and cannot get through the filter paper (1).

(c) $CoCO_3(s) + 2HCl(aq)$ (1) $\rightarrow CoCl_2(aq) + H_2O(l) + CO_2(g)$ (1)

(d) Cobalt metal does not react with dilute acid (1).

8 (a) protons and neutrons in a central nucleus (1); electrons circling nucleus (1)

(b) (i) $^{7}_{3}Li$ has one more neutron than $^{6}_{3}Li$ OR they have different numbers of neutrons (1).

(ii) They have the same number of electrons (1), and the chemical reactions depend on the electronic structure (1).

(c) The outer electrons (1) of lithium atoms are freely moving/delocalised (1) and conduct electricity.

9 (a) The lattice structure has billions of strong covalent bonds, which are hard to break (1), and so it is hard (1).

(b) The fluoride might harm some people (1) and everybody needs the water supply (1).

(c) (i) H^+ ion (1)

(ii) H_2O (1)

Additional Science Physics P2 practice paper

1 (a) $W = m \times g = 75 \times 10$ (1) $= 750$ (1) N

(b) $W = F \times d = 750 \times 4$ (1) $= 3000$ (1) J.

If you got part (a) wrong, but used your answer to part (a) correctly in part (b), you would still get the 2 marks.

(c) (i) 25 J (1), as all the potential energy will be converted into kinetic energy by the time it reaches the ground (1).

(ii) $v^2 = 2 \times E_k/m = 2 \times 25/0.5$ (1) $= 100$ (1) $(m/s)^2$, $v = 10$ (1) m/s

2 (a) Any two from: wear gloves (1), handle source using tongs (1), point source away from body (1), store source in lead-lined box when not in use (1).

(b) (i) Any two from: rocks (1), cosmic rays (1), nuclear accidents (1), hospitals/medical (1), food and drink (1), radon gas (1).

(ii) 5 minutes $= 5 \times 60 = 300$ s (1), average count rate $= 150/300 = 0.5$ (1) counts per second

(c) Beta and gamma (1). Any three from: if alpha radiation was being emitted, the count rate would fall when paper was put in the way (1) because alpha particles are absorbed by a sheet of paper (1). There is some beta radiation, because this is absorbed by aluminium (1), and the count rate falls as thicker aluminium sheets are used (1). There is also gamma radiation, as this can penetrate all the materials used in this test (1).

(d) (i) Points plotted correctly (1), smooth line drawn through points (1).

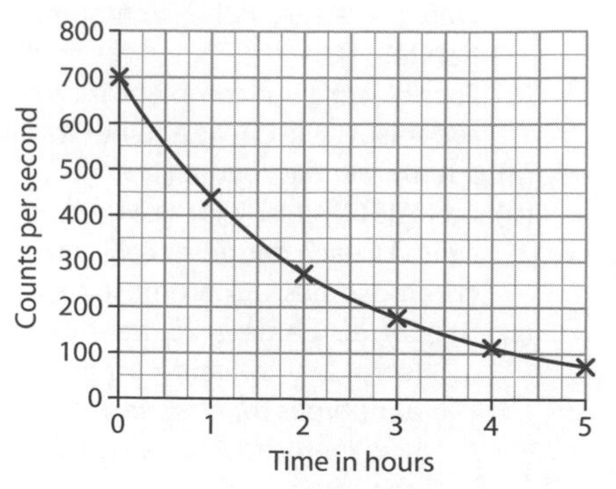

(ii) half-life = 1.5 hours (2) +/− 0.1 hours (1 mark if you got the answer wrong but showed some correct working, such as data taken from the graph, or lines marked on the graph at two count rates, one twice as big as the other)

3 (a) 4 (1) V

(b) $R = V/I = 4/1.6$ (1) $= 2.5$ (1) Ω (1).

If you got part (a) wrong, but used your answer to part (a) correctly in part (b), you would still get the 2 marks.

(c) $P = V \times I = 4 \times 1.6$ (1) $= 6.4$ (1) W.

If you got part (a) wrong, but used your answer to part (a) correctly in part (c), you would still get 2 marks.

(d) (i) 3 V (1)

(ii) current through one bulb $= V/R = 3/4$ (1) $= 0.75$ A (1) current through ammeter $= 3 \times 0.75$ A $= 2.25$ (1) A

4 (a) Distance = area under graph $= \frac{1}{2} \times 12 \times 5$ (1) $=$ 30 (1) m

(b) $E_k = \frac{1}{2} \times m \times v^2 = 0.5 \times 1500 \times 12 \times 12$ (1) $=$ 108 000 (1) J

(c) Work = kinetic energy (1), $F = W/d = 108\,000/30$ (1) $= 3600$ (1) N.

If you got parts (a) or (b) wrong, but used your answers correctly in part (c), you would still get the 2 marks.

(d) A basic answer includes a brief reference to stopping distance being affected by mass, or the stopping distance increasing with increasing speed.

A good answer includes some explanation of why stopping distance increases with mass and/or with speed.

An excellent answer includes a clear, balanced and detailed explanation of why the stopping distance depends on both mass and speed, and links this to the need for different speed limits.

Examples of points made in the response:

- When a car stops, the brakes transfer the kinetic energy to heat up the brakes and the surroundings.
- The greater the mass of a vehicle, the greater its kinetic energy for a given speed.
- The car's brakes provide the same stopping force with and without the caravan.
- So if towing a caravan, the car will take longer/cover a greater distance when coming to a halt from a given speed.
- The higher the speed of the vehicle the longer the stopping distance, as higher speeds mean more kinetic energy to be transferred.
- Therefore, if the car with a caravan is to be able to stop in a safe distance, it must be travelling more slowly than the car without a caravan.

5 (a) $p = m \times v$; $= 13\ \text{kg} \times 2\ \text{m/s}$ (1) $= 26$ (1) kg m/s (1)

(b) momentum of astronaut + momentum of tool bag = 0; momentum of astronaut = −momentum of tool bag (1); $v = p/m = -26\ \text{kg m/s}/65\ \text{kg}$ (1) = −0.4 (1) m/s

6 C (1); the extension of a spring depends on the force divided by the spring constant/*k* (1). If *k* is smaller, then for a given force the extension will be bigger (1).

7 (a) ${}^{214}_{83}\text{Bi} \rightarrow {}^{214}_{84}\text{Po} + {}^{0}_{-1}\text{e}$

(1 for each one correct)

(b) Alpha decay (1), because lead-210 has a mass number 4 less than the polonium/because beta decay does not change the mass number (1).

(c) Atomic number of lead = atomic number of polonium − atomic number of alpha particle = 84 − 2 (1) = 82 (1).

Published by Pearson Education Limited, Edinburgh Gate, Harlow, Essex, CM20 2JE.

www.pearsonschoolsandfecolleges.co.uk

Copies of official specifications for all AQA qualifications may be found on the AQA website: www.aqa.org.uk

Edited by Jim Newall and Florence Production Ltd
Typeset and illustrated by Tech-Set Ltd, Gateshead
Cover illustration by Miriam Sturdee

First published 2013

17 16 15 14 13
10 9 8 7 6 5 4 3 2 1

British Library Cataloguing in Publication Data
A catalogue record for this book is available from the British Library

ISBN 978 1 447 94247 4

Printed in Slovakia by Neografia

Biology
Figures
Graph, page 28, *Maternal Age Effect*, 2006, Wikipedia, http://en.wikipedia.org/wiki/File:Maternal_Age_Effect.png, granted under the GNU Free Documentation License (GFDL).

In the writing of this book, no AQA examiners authored sections relevant to examination papers for which they have responsibility.